THE NUCLEAR OVERHAUSER EFFECT

CHEMICAL APPLICATIONS

The Nuclear Overhauser Effect

CHEMICAL APPLICATIONS

JOSEPH H. NOGGLE
Department of Chemistry
University of Delaware
Newark, Delaware

ROGER E. SCHIRMER
Eli Lilly and Company
Indianapolis, Indiana

ACADEMIC PRESS   New York   San Francisco   London   1971
A Subsidiary of Harcourt Brace Jovanovich, Publishers

ACADEMIC PRESS, INC.
111 Fifth Avenue, New York, New York 10003

United Kingdom Edition published by
ACADEMIC PRESS, INC. (LONDON) LTD.
24/28 Oval Road, London NW1 7DD

LIBRARY OF CONGRESS CATALOG CARD NUMBER: 70-159627

PRINTED IN THE UNITED STATES OF AMERICA

79 80 81 82   9 8 7 6 5 4 3 2 1

CONTENTS

CHAPTER 3. The Nuclear Overhauser Effect in Rigid Molecules

CHAPTER 4. The Effects of Internal Motions

CHAPTER 5. Experimental Methods

CHAPTER 6. Transient Methods

CHAPTER 7. The Effects of Chemical Exchange

CHAPTER 8. Applications of the Nuclear Overhauser Effect: A Review of the Literature

APPENDIX I. Tightly Coupled Spins   226

APPENDIX II. Mathematical Methods

Bibliography   244

PREFACE

The nuclear Overhauser effect (NOE), which is the change in the integrated intensity of the NMR absorption of a nuclear spin as a result of the concurrent saturation of another NMR resonance, has found limited use over the past two decades in the study of chemical kinetics and, somewhat more recently, in the assignment of NMR spectra. Lately, interest in the NOE has grown enormously following the realization that detailed qualitative and quantitative information on molecular configuration and conformation can be obtained from it. The uniqueness of this approach to problems in molecular structure, together with the increasing availability of NMR spectrometers sufficiently sophisticated for NOE studies and sufficiently simple in operation to be used on a routine basis, have increased and will continue to increase the applications of the method manyfold. However, the numerous existing books on NMR written primarily for chemists barely mention the NOE and do not provide the background in nuclear relaxation theory necessary to understand it. Aside from original research papers, only a few references of a highly theoretical nature are extant. The difficulty and rigor of these references has been a source of frequent misunderstandings and has surely limited the growth of the field.

Our purpose in writing this monograph was to provide a unified treatment of both the theory and applications of the NOE that would be accessible to the chemist who is not a specialist in NMR, but who would nevertheless like to use the NOE to solve problems in his own area of interest. Assuming only that the reader has a typical chemist's working knowledge of NMR, we have attempted to present the results in each section in a sufficiently simple form so that little mathematical sophistication will be required to apply them with understanding. Indeed, in many instances, completeness and rigor have been deliberately sacrificed to this end. On the other hand, enough of the theory has been presented to provide a reasonably complete foundation for the chemist who is more physically oriented and who would like to understand the subject in greater depth.

The theory of NMR relaxation required in the discussion of the nuclear Overhauser effect is covered in Chapters 1 and 2. In Chapter 3, the basic equation describing the NOE in rigid, multispin molecules is developed, and explicit expressions are presented for the quantitative evaluation of NOE data in the special cases that are most likely to be of practical importance. The theory is extended to nonrigid molecules in Chapter 4, and experimental considerations are reviewed briefly in Chapter 5. The theory of transient experiments, which often prove to be useful supplements to the more common steady state NOE experiments, is presented in Chapter 6. A brief discussion of the Fourier transform method is also included in Chapter 6. Chapter 7 contains the extension of the theory to systems of exchanging nuclei, with emphasis on the use of the NOE to study the exchange process itself: A review of the applications of the NOE to exchanging systems that have been made is also included in this chapter. The remainder of the literature up to December, 1970 is reviewed in Chapter 8. The Bibliography at the end of the book covers the literature through June, 1971.

ACKNOWLEDGMENTS

We wish to acknowledge the support of Eli Lilly and Company and the Chemistry Department of the University of Wisconsin during the preparation of this manuscript. We are also grateful to the many individuals who generously supplied figures to be used in the text, and to the journals that granted permission to reproduce material from their pages. We are particularly indebted to our co-workers, Dr. Phillip A. Hart and Jeffrey P. Davis for their many helpful suggestions and comments and for the fruitful collaboration we have had with them on some of the research described herein.

INTRODUCTION

The term "Overhauser effect" referred originally to the dynamic polarization of nuclei in metals (*1*) when the spin resonance of the electrons was saturated. The first application of this effect in a system containing only nuclear spins was made by Solomon and Bloembergen (*2*) in their study of chemical exchange in HF. The nuclear Overhauser effect (NOE) next found application in the assignment of complex nuclear magnetic resonance (NMR) spectra (*3*), in the study of chemical exchange (*4*) and nuclear relaxation (*5*), and in signal-to-noise improvement in NMR spectra (*6*). The potential of the NOE for providing information on the conformation and configuration of molecules in solution was first demonstrated by Anet and Bourn (*7*), and since that time applications in this area have grown rapidly. Recently, Bell and Saunders (*8*) reported direct correlation between NOE enhancements and internuclear distances, and Schirmer, Noggle, Davis, and Hart (*9*) have demonstrated that relative internuclear distances can be determined quantitatively from NOE measurements on systems containing three or more spins.

The latter paper in particular shows that an understanding of nuclear spin–lattice relaxation allows much more information to be obtained from an NOE experiment than is otherwise available. Numerous books explaining the principles and practices of high-resolution NMR in relatively simple terms

1

have appeared in recent years. Most of these books, however, have neglected the important area of spin–lattice relaxation so that the chemist who wishes to understand and use the NOE has only the literature and reference volumes such as Abragam's "Principles of Nuclear Magnetism" (*10*) as sources. It is this lack of an intermediate level treatment of the subject which we hope to remedy.

We shall assume that the reader is familiar with the routine operation of commercial NMR spectrometers and with the theory of NMR at the level of the books by Bovey (*11*) or Becker (*12*). We shall be primarily concerned with the use of steady-state NOE and supplementary transient techniques for the study of molecular structure and conformation, and chemical exchange. The NOE is sufficiently simple that very little mathematical sophistication is required. The reader who is not interested may ignore the few derivations presented in the text. Since the NOE is a nuclear magnetic double resonance (NMDR) effect, the reader may find helpful, but not necessary, one of the review articles by Hoffman and Forsén (*13*), Baldeschwieler and Randall (*14*), or McFarlane (*15*). For the most part we will be concerned with situations where the more esoteric effects of NMDR can be avoided; to wit, loosely coupled spins ($J \ll \delta$) in the strong rf or decoupling limit ($\gamma H_2 \gg J$). In addition, when spin multiplets due to J coupling are present, it will be assumed that all components of the NMR resonance of the irradiated spins are saturated by the strong rf and that the "intensity" of the enhanced spin refers to the total integrated intensity of all the NMR lines belonging to the detected spin. The many interesting effects which occur when these simplifying conditions are not used have been reviewed by Nageswara Rao (*16*).

Many of the statements herein may not apply in all generality; this is because we have attempted to find a compromise between generality and simplicity which will be adequate in all cases where the nuclear Overhauser effect is likely to be of interest. Those who wish a better understanding of the fundamentals of nuclear relaxation may find help in the monographs by Carrington and McLachlan (*17*) and Slichter (*18*).

Common symbols which are more or less standard will often be used without explanation. Some of these are:

$\hbar$:	Planck's constant divided by 2π
k:	Boltzmann's constant
T:	the absolute temperature (degrees kelvin)
J:	the scalar spin–spin coupling constant (hertz)
δ:	the relative chemical shift (hertz)
$\omega_0, \omega_1, \omega_S, \omega_i$:	Larmor frequencies (radians per second)
H_0, H_1, H_2:	the intensities of the dc magnetic field, the observing rf field and the saturating rf field, respectively

α, β: Spin-$\frac{1}{2}$ wave functions—eigenfunctions of $\mathbf{I}_z$.

I, S, I_i: the total spin quantum numbers. The magnitude of the angular momentum of spin $\mathbf{I}$ is $[I(I+1)]^{1/2}\hbar$ and the magnetic moment is $\gamma\hbar I$.

γ: the gyromagnetic ratio

REFERENCES

1. A. W. Overhauser, *Phys. Rev.* **92**, 411 (1953).
2. I. Solomon and N. Bloembergen, *J. Chem. Phys.* **25**, 261 (1956); another early paper on the NOE was that of J. Wirtz, P. L. Jain and R. L. Batdorf, *Phys. Rev.* **102**, 920 (1956).
3. R. Kaiser, *J. Chem. Phys.* **39**, 2435 (1963).
4. S. Forsén and R. A. Hoffman, *J. Chem. Phys.* **39**, 2892 (1963).
5. J. Noggle, *J. Chem. Phys.* **43**, 3304 (1965).
6. P. C. Lauterbur, quoted in reference *14*.
7. F. A. L. Anet and A. J. R. Bourn, *J. Amer. Chem. Soc.* **87**, 5250 (1965).
8. R. A. Bell and J. K. Saunders, *Can. J. Chem.* **48**, 1114 (1970).
9. R. E. Schirmer, J. H. Noggle, J. P. Davis, and P. A. Hart, *J. Amer. Chem. Soc.* **92**, 3266 (1970); erratum, **92**, 7239 (1970).
10. A. Abragam, "The Principles of Nuclear Magnetism," Oxford Univ. Press, London and New York, 1961.
11. F. A. Bovey, "Nuclear Magnetic Resonance Spectroscopy," Academic Press, New York, 1969.
12. E. D. Becker, "High Resolution NMR," Academic Press, New York, 1969.
13. R. A. Hoffman and S. Forsén, *Progr. NMR Spectros.* **1**, 15 (1966).
14. J. D. Baldeschwieler and E. W. Randall, *Chem. Rev.* **63**, 81 (1963).
15. W. McFarlane, *Annu. Rev. NMR Spectros.* **1**, 135 (1968).
16. B. D. Nageswara Rao, *Advan. Magn. Resonance* **4**, 271 (1970).
17. A. Carrington and A. D. McLachlan, "Introduction to Magnetic Resonance," Harper, New York, 1967.
18. C. P. Slichter, "Principles of Magnetic Resonance," Harper, New York, 1963.

NUCLEAR SPIN–LATTICE RELAXATION

A. The Nuclear Overhauser Effect

The nuclear Overhauser effect (NOE) is a change in the integrated nuclear magnetic resonance (NMR) absorption intensity of a nuclear spin when the NMR absorption of another spin is saturated. The spins involved may be either heteronuclear or chemically shifted homonuclear spins. Before considering how this phenomenon might occur, let us examine the effect of an rf field on a single nuclear spin.

A spin-$\frac{1}{2}$ nucleus in a magnetic field of intensity H_0 will have two energy levels, customarily labeled α and β, which differ in energy by

$$\Delta E = \gamma \hbar H_0$$

If an rf field with frequency equal to the Larmor frequency of the nuclear spins,

$$v = \Delta E/h = (\gamma/2\pi) H_0$$

is applied to these spins, transitions will be induced between the two energy levels. The number of transitions $\alpha \to \beta$ caused by the rf field will be proportional to P_α, the population of the α state; the number of transitions $\beta \to \alpha$ will similarly be proportional to P_β. The net absorption intensity is therefore proportional to the difference in populations, $P_\beta - P_\alpha$. Initially, of course, the lower energy level will have a larger population so that the number of upward transitions will exceed the number of downward transitions. If the rf field is very strong, this process will result in *saturation*; that is, the populations of the two energy levels will become equal and absorption of rf energy will cease.

Some mechanisms must exist for these populations to return to equilibrium. In optical spectroscopy this process would occur by spontaneous emission; however, at the frequencies common in NMR spontaneous emission is a very inefficient mechanism. The manner by which nuclear spins return to equilibrium is called *spin–lattice relaxation*. We shall discuss this subject in more detail later; here we shall simply define a quantity $W_{\alpha\beta}$ as the probability per unit time that a transition will occur between the states α and β due to spin–lattice relaxation. A more detailed discussion of rf absorption including the effect of spin–lattice relaxation is given by Carrington and McLachlan (*1*).†

We shall now discuss two nuclear spins-$\frac{1}{2}$ assuming, initially, that they are of the same nuclear species, chemically shifted but *not J* coupled. We shall call these spins I and S and the energy levels will be

level 1   spin I is α, spin S is α—$\alpha\alpha$

level 2   spin I is α, spin S is β—$\alpha\beta$

level 3   spin I is β, spin S is α—$\beta\alpha$

level 4   spin I is β, spin S is β—$\beta\beta$

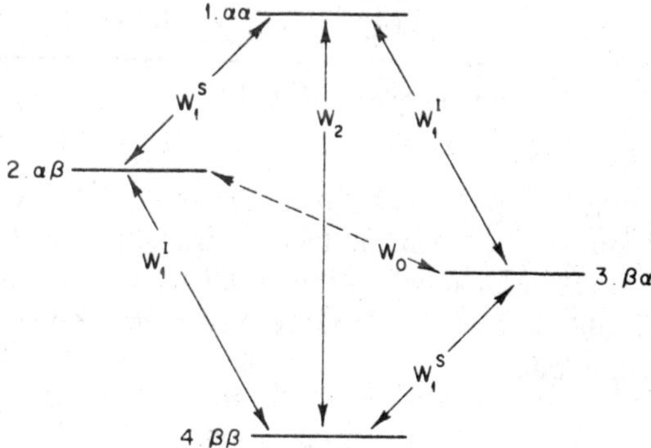

Fig. 1-1 Energy level diagram for two spins-$\frac{1}{2}$. The W's are spin–lattice transition probabilities. The state of spin I is listed first; e.g., $\alpha\beta$ means spin I is α and spin S is β.

---

† Note that the transition probability W used by Carrington and McLachlan (*1*) is different from that used herein; specifically in Reference *1*, $W_{\alpha\beta} \neq W_{\beta\alpha}$.

These levels are shown schematically in Fig. 1-1. Description of the spin–lattice relaxation requires four transition probabilities:

(1) W_1^{I}; the single quantum transition probability that spin I will go from α to β (or β to α) while the state of spin S remains unchanged.

(2) W_1^{S}; the single quantum transition probability for spin S when spin I remains unchanged.

(3) W_2; the two quantum transition probability for the two spins to relax simultaneously in the same direction, i.e., $\alpha\alpha \rightarrow \beta\beta$ or $\beta\beta \rightarrow \alpha\alpha$.

(4) W_0; the zero quantum transition probability for a mutual spin flip, $\alpha\beta \rightarrow \beta\alpha$ or $\beta\alpha \rightarrow \alpha\beta$.

The equilibrium populations (denoted here P°) of levels 2 and 3 will be nearly equal since the energies of these levels are nearly equal; we shall denote these populations as Π. The population of level 1 will be smaller by an amount which we shall call δ'; the population of level 4 will be larger than Π by an equal amount. The equilibrium populations are given in Table 1-1.

TABLE 1-1

Level	Equilibrium population	Spin S is saturated	Effect of W_2 alone[a]	Effect of W_0 alone[a]
1. $\alpha\alpha$	$\Pi-\delta'$	$\Pi-\tfrac{1}{2}\delta'$	$\Pi-\tfrac{1}{2}\delta'-d$	nc
2. $\alpha\beta$	Π	$\Pi-\tfrac{1}{2}\delta'$	nc	$\Pi-\tfrac{1}{2}\delta'+d$
3. $\beta\alpha$	Π	$\Pi+\tfrac{1}{2}\delta'$	nc	$\Pi+\tfrac{1}{2}\delta'-d$
4. $\beta\beta$	$\Pi+\delta'$	$\Pi+\tfrac{1}{2}\delta'$	$\Pi+\tfrac{1}{2}\delta'+d$	nc

[a] nc means no change from the value given in Column 3.

The NMR resonance of spin S will consist of two transitions, 1–2 and 3–4; in the absence of J coupling between I and S these transitions will have the same frequency. Likewise the resonance of spin I will have two components, the 1–3 and 2–4 transitions. The absorption intensity of the I spins will therefore be proportional to the quantity

$$(P_3 - P_1) + (P_4 - P_2)$$

whose equilibrium value can be seen to be $2\delta'$ (Tables 1-1 and 1-2). If a strong rf field is applied at the resonance frequency of spin S, that spin will be saturated with the result that

$$P_1 = P_2 \quad \text{and} \quad P_3 = P_4$$

Since the total number of spins

$$P_1 + P_2 + P_3 + P_4 = 4\Pi$$

must remain unchanged, one can easily show that the populations which must result from the saturation of spin S are those in Column 3 of Table 1-1. These population changes, however, do not result in any change in the I spin resonance absorption intensity as can be seen in Table 1-2.

TABLE 1-2

	$P_3 - P_1$	$P_4 - P_2$	Intensity of the spin I resonance
Equilibrium	δ'	δ'	$2\delta'$
Saturate S	δ'	δ'	$2\delta'$
W_1^{I} alone	δ'	δ'	$2\delta'$
W_1^{S} alone	δ'	δ'	$2\delta'$
W_2 alone	$\delta' + d$	$\delta' + d$	$2\delta' + 2d$
W_0 alone	$\delta' - d$	$\delta' - d$	$2\delta' - 2d$

We now consider, one at a time, the effects of the spin–lattice transition probabilities. As can be seen in Fig. 1-1, W_1^{I} will cause relaxation between the levels 1–3 and 2–4. However, the saturation of spin S has left these energy levels at equilibrium with each other, that is,

$$P_1 - P_3 = P_1{}^\circ - P_3{}^\circ = \delta'$$

and

$$P_2 - P_4 = P_2{}^\circ - P_4{}^\circ = \delta'$$

(cf. Table 1-2) and therefore there will be no net change in the populations due to W_1^{I} alone.

W_1^{S} causes relaxation between the 1–2 and 3–4 levels. But these are the levels whose population the strong rf is requiring to be equal; W_1^{S} will be ineffective in altering the populations so long as the strong rf is present.

On the other hand, the population difference $P_4 - P_1 = \delta'$ is less than the equilibrium value $P_4{}^\circ - P_1{}^\circ = 2\delta'$. The effect of W_2 will be to attempt to re-establish equilibrium by increasing P_4 and decreasing P_1. If the amount of population transferred from level 1 to level 4 is d, we will have

$$P_3 - P_1 = P_4 - P_2 = \delta' + d$$

and the absorption intensity of spin I will be increased by $2d$ over its equilibrium value (cf. Table 1-2).

As can be seen in Column 3 of Table 1-2, saturation of S causes $P_3 - P_2 = \delta'$ whereas at equilibrium these populations were equal. The effect of W_0 will therefore be to increase P_2 and decrease P_3. If the amount of population transferred from level 3 to level 2 by spin–lattice relaxation is again defined as d, the effect (Table 1-2) will be to decrease the intensity of the NMR resonance of spin I.

In a realistic case, of course, all of the transition probabilities will be simultaneously effective; the resulting compromise will be derived later.

If the spins I and S are J coupled, each resonance will be a doublet. If only one of the S transitions were saturated intensity changes [as well as additional multiplet splittings due to double resonance effects (2)] could occur even in the absence of spin–lattice relaxation; Kaiser (3) has analyzed such effects in multispin systems using a method similar to that used in the preceding paragraphs. Hoffman and Forsén (2) have suggested the term "generalized Overhauser effects" for changes in intensities which are not related to spin–lattice relaxation or chemical exchange. These effects are of no interest in the present work.

The preceding analysis for uncoupled spins will apply to J coupled spins if: (a) the spins are loosely coupled, $J_{IS} \ll \delta_{IS}$ (some discussion of tightly coupled spins and the restriction on the J/δ ratio will be given in Appendix I); (b) all lines of the S resonance are saturated; (c) the *total* intensity of the I resonance is measured.

Before discussing the NOE further, we must consider the topic of nuclear relaxation.

B. Relaxation Times

While the preceding analysis in terms of populations and transition probabilities is more suited to our principal topic, nuclear relaxation is more frequently discussed in terms of the macroscopic magnetization of the nuclear spins and the relaxation times T_1 and T_2.

When nuclear spins are placed in a magnetic field, a small but observable magnetization M_z will be induced due to the partial alignment of the spins along the direction of the magnetic field. After the spins have been in the field for a time, they will reach thermal equilibrium with their surroundings and M_z will assume its equilibrium value M_0; the approach of M_z to its equilibrium value is given by

$$dM_z/dt = -(1/T_1)(M_z - M_0) \tag{1.1}$$

This equation applies equally well to the return of M_z to equilibrium with the spins already in the magnetic field; M_z could be removed from its equilibrium value, for example, by applying a strong rf field at the spin's Larmor frequency. Equation (1.1), always valid for an ensemble of identical spins whose motions are uncorrelated, serves to define the spin–lattice relaxation time T_1, which is a measure of the time required for the nuclear spins to dissipate their excess energy to their surroundings. By surroundings, or *lattice*, we usually mean the translational or rotational degrees of freedom of the molecules in which the nuclei are situated. The exact means by which energy is transferred between the spins and the lattice is called the *relaxation mechanism* and will be discussed in detail in Chapter 2.

The spin–lattice relaxation time T_1 must be sharply distinguished from the spin–spin relaxation time T_2. If an rf magnetic field is applied at the spin's Larmor frequency, it will induce a transverse magnetization M_x or M_y which precesses at the Larmor frequency about the direction of the dc magnetic field H_0. If the rf field is removed, the spins continue to precess but will become dephased with the result that the macroscopic transverse magnetization will disappear; this process may or may not involve a transfer of energy between the spins and lattice. The equation

$$M_x(t) = M_x(0) \exp(i\omega_0 t) \exp(-t/T_2) \tag{1.2}$$

serves to define T_2 where ω_0 is the Larmor frequency of the spins and $M_x(0)$ the value of M_x at $t = 0$. The equilibrium value is, of course,

$$M_x(t \to \infty) = 0$$

More commonly, T_2 is related to the linewidth of an unsaturated NMR absorption in the absence of magnetic field inhomogeneities or instrumental instability by

$$\Delta = 1/\pi T_2 \tag{1.3}$$

where Δ is the full width of the absorption line at half-maximum intensity in units of hertz (i.e., cycles per second). Equation (1.3) is an equally good definition of T_2, but, since field inhomogeneities are commonly the major cause of linewidth for spin-$\frac{1}{2}$ nuclei, it is not particularly useful for measuring T_2 of spin-$\frac{1}{2}$ nuclei.

By analogy with Eq. (1.3) one occasionally defines an "effective T_2" or T_2^* by

$$T_2^* = 1/\pi\Delta_{\text{obs}} \tag{1.4}$$

where Δ_{obs} is the measured linewidth including field inhomogeneities. The concept of an effective T_2, while lacking a great deal in exactness, is frequently useful. There is no simple relationship between T_2 and T_2^*.

Spin–lattice relaxation can, of course, contribute to T_2; in fact

$$1/T_2 = \tfrac{1}{2}(1/T_1) + 1/T_2' \tag{1.5}$$

where T_2' represents physical processes which dephase the spins without exchanging energy with the lattice. One such process is the adiabatic exchange of energy among the spins—hence the name spin–spin relaxation. The fact that T_1 and T_2 are often numerically equal to each other should not be interpreted to mean that they are identical concepts. Since it is T_1 which is of greatest importance to the theory of the NOE, we shall say little else about T_2.

Equations (1.1) and (1.2) are two of the Bloch equations† written for the case when rf fields are absent. One might think at first that the full set of Bloch equations including terms due to the rf field would be needed to describe high-resolution NMR experiments since rf fields are always present during such experiments. We shall avoid these complications by making one or the other of two limiting assumptions:

(a) The rf field is very weak so that the value of M_z (not necessarily equal to M_0) is not changed when the spectrum is observed. Under these conditions M_z will be proportional to the total integrated intensity (i.e., area) of the NMR resonance including all components of multiplets due to scalar or J coupling if present.

(b) The rf field is very strong so that the resonance, including all components of J-coupled multiplets, is saturated. Under these conditions the M_z of the saturated spins is equal to zero.

C. The Measurement of T_1

It may help illustrate the concept of spin–lattice relaxation if we consider the manner in which T_1 is measured. A simple technique (4) which can be used on most high-resolution NMR spectrometers without extensive modification will be described although more accurate methods are available (5). This technique is useful if (a) the resonance whose T_1 is to be measured is well separated from the other lines of the spectrum; (b) $T_1 \gtrsim 5$ sec; (c) the signal-to-noise ratio is very good; and (d) the J couplings are small enough so that a single line can be observed under low resolution. The technique to be described is one of several adiabatic rapid passage (ARP) techniques.

---

† The Bloch equations are treated in most books on NMR including References *1*, *5*, and *7* (Chap. III) of this chapter and References *11* and *18* of the Introduction.

The NMR signal of the spins whose T_1 is to be measured is displayed on an oscilloscope with the spectrometers saw-tooth sweep unit (period on the order of several seconds). The rf field intensity is reduced until the spins are not saturated and M_z is not altered by the measurement. The repeated peaks can be conveniently recorded on a strip-chart (x–t) recorder. The heights of the peaks should not be altered if the sweep width or frequency is changed; if such an effect occurs, it indicates either (a) saturation, (b) excessive filtering, or (c) that the response of the recorder is too slow.

Then, for the duration of a single sweep, the rf amplitude is increased by 40–60 dB; if properly done, this sweep will be an adiabatic passage and, immediately following this sweep, the magnetization of the spins will be inverted. If time zero is the time of the ARP, $M_z(t = 0) = -M_0$. According to Eq. (1.1) the value of M_z measured at time t following the passage will be

$$\ln\left[M_0 - M_z(t)\right] = \ln 2M_0 - t/T_1 \tag{1.6}$$

$M_z(t)$, taken to be proportional to the signal intensity at time t, is measured by the weak rf each time the saw-tooth sweep passes through resonance.

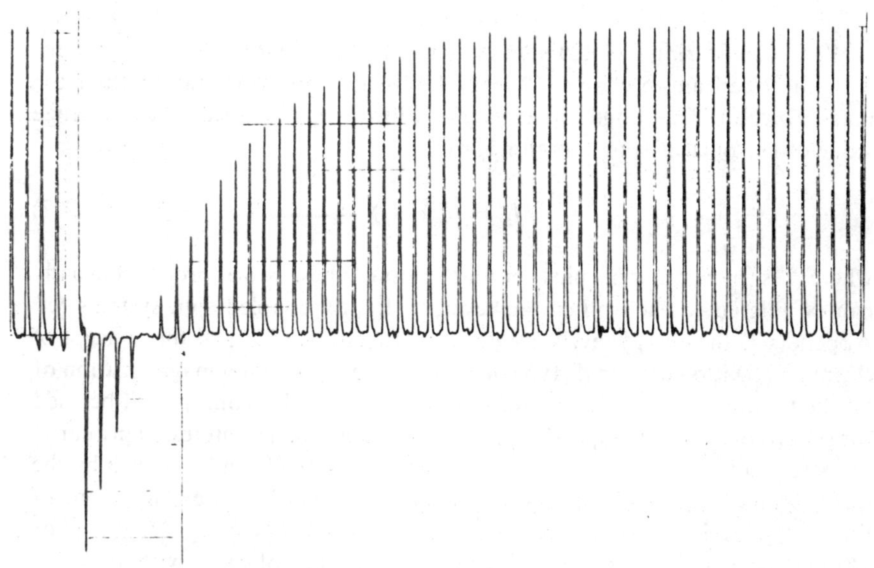

Fig. 1-2 An adiabatic rapid passage measurement of the T_1 of the protons of 2,3-dibromothiophene. The spins were inverted by an ARP at the left (where the recorder pen is off scale) and are seen recovering to their equilibrium intensity. Each peak includes all four lines of the unresolved AB quartet of the two protons. Successive peaks are about 4 sec apart and $T_1 = 29$ sec.

In practice, extrapolations of $M_z(t)$ to $t = 0$ rarely go to $-M_0$, presumably because the adiabatic conditions were not properly fulfilled. However, if $\log(M_0 - M_z)$ is plotted versus t, the slope will give a good value of T_1 which is independent of the precise value of $M_z(0)$ or the location of the origin, $t = 0$. Other methods, including the method of half-lives, will be inaccurate.

An example (6) of the use of ARP to measure the T_1 of the protons of 2,3-dibromothiophene is shown in Fig. 1-2; $T_1 = 29$ sec.

Despite the crudeness of this technique and the fact that it presumes Eq. (1.1) is valid, values of T_1 obtained in this manner can provide valuable supplementary information to the steady-state NOE measurements.

Important experimental parameters whose effect must be considered are (a) rf field strength for observation and the ARP, (b) sweep frequency, (c) sweep width, and (d) filtering. Intensity measurements are more reproducible if the magnetic field is not too homogeneous (not too much "ringing," if any). The first peak following the ARP usually must be discarded. The "infinity" value of the intensity, proportional to M_0, must be measured carefully.

D. Relaxation of Coupled Spins

When considering coupled spin systems, it is frequently necessary to work with the quantum-mechanical energy levels and operators rather than the magnetization. The longitudinal magnetization M_z is related to the average value of the nuclear spin operator $\mathbf{I}_z$ by

$$M_z = N\gamma\hbar \langle \mathbf{I}_z \rangle \tag{1.7}$$

where N is the number of spins per unit volume. These two approaches usually can be related to each other; however, in tightly coupled spin systems (cf. Appendix I) the energy levels are not eigenfunctions of $\mathbf{I}_z$ and the concept of the magnetization of a single type of spin is useless. The total magnetization of all the tightly coupled spins is still well defined, and this concept will be used for groups of equivalent spins, e.g., methyl groups. We are interested primarily in loosely coupled spins; in such cases each energy level can be labeled by the quantum number M_k of the $\mathbf{I}_z(k)$ of each nuclear spin k. For example, one of the energy levels of a three spin system is ($M_1 = +\frac{1}{2}$, $M_2 = -\frac{1}{2}$, $M_3 = +\frac{1}{2}$) or $\alpha\beta\alpha$ in the conventional notation where $\alpha(k)$ is the state of spin k with $M_k = +\frac{1}{2}$ and $\beta(k)$ is the state of spin k with $M_k = -\frac{1}{2}$.

Each of the energy levels i of a coupled spin system has a population P_i whose value at equilibrium is given by the Boltzmann equation

$$P_i^\circ = K \exp(-E_i/kT) \tag{1.8}$$

where E_i is the energy of the level i and K is a normalization constant. The approach of P_i to equilibrium is given by the **master equation for populations** (7)

$$dP_i/dt = \sum_j W_{ij}(P_j - P_j^\circ) - (P_i - P_i^\circ) \sum_j W_{ij} \tag{1.9}$$

where $W_{ij} = W_{ji}$ is the transition probability per unit time for a transition between levels i and j caused by spin–lattice relaxation.

For loosely coupled spins, the magnetization of a given spin k is given by Eq. (1.7) and

$$\langle I_z(k) \rangle = \sum_i M_k^i P_i \tag{1.10}$$

where M_k^i is the quantum number of $I_z(k)$ appropriate to level i and the sum is over all energy levels (n nuclear spins-$\frac{1}{2}$ have 2^n energy levels).

As an illustrative example consider a single type of nuclear spin with energy levels α (population P_α) and β (population P_β). Then

$$\langle I_z \rangle = \tfrac{1}{2}P_\alpha - \tfrac{1}{2}P_\beta$$

$$M_z = \tfrac{1}{2}N\gamma\hbar(P_\alpha - P_\beta) \quad \text{and} \quad M_0 = \tfrac{1}{2}N\gamma\hbar(P_\alpha^\circ - P_\beta^\circ)$$

Equation (1.9) is now

$$dP_\alpha/dt = W_{\alpha\beta}(P_\beta - P_\beta^\circ) - W_{\alpha\beta}(P_\alpha - P_\alpha^\circ)$$

and

$$dM_z/dt = -2W_{\alpha\beta}(M_z - M_0)$$

When this is compared to Eq. (1.1), we see that for this special case $1/T_1 = 2W_{\alpha\beta}$.

E. Relaxation of Two Spins

We will now consider Eq. (1.9) in detail for a system of two loosely coupled spins in order to obtain a simpler equation [analogous to Eq. (1.1)] for the magnetization.

We have two spins-$\frac{1}{2}$, I and S, with their scalar coupling constant $J_{IS} \ll \delta_{IS}$, their relative chemical shift. There will be four energy levels which will be labeled as before (cf. Fig. 1-1). From Eq. (1.10)

$$\langle I_z \rangle = \tfrac{1}{2}P_1 + \tfrac{1}{2}P_2 - \tfrac{1}{2}P_3 - \tfrac{1}{2}P_4 \tag{1.11a}$$

$$\langle S_z \rangle = \tfrac{1}{2}P_1 - \tfrac{1}{2}P_2 + \tfrac{1}{2}P_3 - \tfrac{1}{2}P_4 \tag{1.11b}$$

Also from Eq. (1.9)

$$dP_1/dt = -(W_{12} + W_{13} + W_{14})(P_1 - P_1^{\,0}) + W_{12}(P_2 - P_2^{\,0})$$
$$+ W_{13}(P_3 - P_3^{\,0}) + W_{14}(P_4 - P_4^{\,0}) \qquad (1.12)$$

and similar expressions for P_2, P_3, and P_4.

With these equations and Eq. (1.11a) we get

$$d\langle \mathbf{I}_z \rangle/dt = -(W_{13} + W_{14})(P_1 - P_1^{\,0}) - (W_{24} + W_{23})(P_2 - P_2^{\,0})$$
$$+ (W_{13} + W_{23})(P_3 - P_3^{\,0}) + (W_{14} + W_{24})(P_4 - P_4^{\,0}) \qquad (1.13)$$

In the loose coupling limit we have the following equalities:

$$W_{13} = W_{24} = W_1^{\mathrm{I}} \qquad (1.14a)$$

the single quantum transition probability of spin I;

$$W_{12} = W_{34} = W_1^{\mathrm{S}} \qquad (1.14b)$$

the single quantum transition probability of spin S;

$$W_{23} = W_0 \qquad \text{and} \qquad W_{14} = W_2 \qquad (1.14c)$$

the zero and two quantum transition probabilities, respectively. We note that the equilibrium value of $\langle \mathbf{I}_z \rangle$ is

$$I_0 = \tfrac{1}{2}P_1^{\,0} + \tfrac{1}{2}P_2^{\,0} - \tfrac{1}{2}P_3^{\,0} - \tfrac{1}{2}P_4^{\,0} \qquad (1.15)$$

and similarly for S_0. Using the equality (1.14a) in Eq. (1.13) gives

$$d\langle \mathbf{I}_z \rangle/dt = -2W_1^{\mathrm{I}}[\langle \mathbf{I}_z \rangle - I_0]$$
$$- W_0[(P_2 - P_2^{\,0}) - (P_3 - P_3^{\,0})] - W_2[(P_1 - P_1^{\,0}) - (P_4 - P_4^{\,0})] \qquad (1.16)$$

so that an equation of the type Eq. (1.1) will not apply unless W_0 and W_2 are zero; this will be true of some relaxation mechanisms, but not of the most interesting mechanism, dipole–dipole.

Equation (1.16) and its analog for S_z can be put into a more convenient form if we note that [Eqs. (1.11)]

$$\langle \mathbf{I}_z \rangle + \langle \mathbf{S}_z \rangle = P_1 - P_4 \qquad (1.17a)$$

$$\langle \mathbf{I}_z \rangle - \langle \mathbf{S}_z \rangle = P_2 - P_3 \qquad (1.17b)$$

and thus

$$d\langle \mathbf{I}_z\rangle/dt = -\rho_I[\langle \mathbf{I}_z\rangle - I_0] - \sigma_{IS}[\langle \mathbf{S}_z\rangle - S_0] \tag{1.18a}$$

$$d\langle \mathbf{S}_z\rangle/dt = -\rho_S[\langle \mathbf{S}_z\rangle - S_0] - \sigma_{SI}[\langle \mathbf{I}_z\rangle - I_0] \tag{1.18b}$$

$$\rho_I = 2W_1^I + W_0 + W_2 \tag{1.19a}$$

$$\rho_S = 2W_1^S + W_0 + W_2 \tag{1.19b}$$

$$\sigma_{IS} = \sigma_{SI} = W_2 - W_0 \tag{1.19c}$$

In Eqs. (1.18) we use the notation of Solomon (8) which is related to that of Abragam (7) by

$$\rho_I = 1/T_1^{II}, \qquad \sigma_{IS} = 1/T_1^{IS}$$

$$\rho_S = 1/T_1^{SS}, \qquad \sigma_{SI} = 1/T_1^{SI}$$

Equations (1.18) are more general than is implied here. According to Abragam (7) they apply if I and S are other than spin-$\frac{1}{2}$; in this more general case σ_{IS} is not equal to σ_{SI} (cf. Chapter 2). The σ terms are often referred to as the *cross-relaxation* terms. In order for these terms to be nonzero, a relaxation mechanism which couples I and S must be present; there are three such mechanisms:

(a) dipole–dipole relaxation between I and S;
(b) I to S scalar (J) coupling modulated by chemical exchange or internal motion (J is time dependent);
(c) I to S scalar (J) coupling modulated by rapid relaxation of S.

Only (a) and (b) will be significant for systems which are likely to be studied by the NOE technique. Other relaxation mechanisms will cause I and S to relax independently and Eqs. (1.18) will degenerate to Eq. (1.1) with $1/T_1 = \rho = 2W_1$. When cross-relaxation is present, T_1 cannot always be defined.

F. The Nuclear Overhauser Effect for Two Spins

It is the cross-relaxation terms in Eqs. (1.18) which make the NOE possible. Consider the following experiment: (a) a strong rf field is applied at the Larmor frequency of spin S so that $\langle \mathbf{S}_z\rangle = 0$; (b) $\langle \mathbf{I}_z\rangle$ is measured by a weak rf field with a frequency near the Larmor frequency of spin I. The fractional enhancement of the integrated intensity of I ($\langle \mathbf{I}_z\rangle$) when S is saturated compared to its equilibrium intensity (I_0) is defined as

$$f_I(S) \equiv (\langle I_z \rangle - I_0)/I_0 \qquad (1.20)$$

When a steady state is reached, $d\langle I_z \rangle/dt = 0$, Eq. (1.18a) gives

$$\langle I_z \rangle = I_0 + \sigma_{IS} S_0/\rho_I \qquad (1.21a)$$

or

$$f_I(S) = \sigma_{IS} S_0/\rho_I I_0 \qquad (1.21b)$$

Given S and I and γ_S and γ_I the total spin quantum numbers and gyromagnetic ratios of S and I, respectively, we have

$$I_0 \propto I(I+1)\gamma_I, \qquad S_0 \propto S(S+1)\gamma_S$$

giving

$$f_I(S) = \gamma_S \sigma_{IS} S(S+1)/\gamma_I \rho_I I(I+1) \qquad (1.22)$$

The γ factors in Eq. (1.22) are very important for heteronuclear NOE; the I and S factors are unimportant since they occur also in σ/ρ and cancel (see below). In most cases of interest $I = S = \frac{1}{2}$; in such cases Eq. (1.22) becomes

$$f_I(S) = \frac{W_2 - W_0}{2W_1^I + W_0 + W_2} \qquad (1.23)$$

The relaxation mechanism of greatest interest with respect to the NOE is the dipole–dipole relaxation mechanism because of its explicit dependence on the distance between I and S. However, if spin I relaxes *only* by dipole–dipole coupling with S then (7, 8)

$$\sigma_{IS}/\rho_I = I(I+1)/2S(S+1) \qquad (1.24)$$

and thus (without assuming $I = S$), sub in eq 1.22

$$f_I(S) = \gamma_S/2\gamma_I \qquad (1.25)$$

a constant which does not depend on the internuclear distance at all. Of course, in any practical case, other relaxation processes will contribute to ρ_I (i.e., to W_1^I) so that Eq. (1.25) represents the *maximum* NOE enhancement observable in any given case. This begs the question of whether any structural information is available at all from the NOE since these "other" mechanisms are frequently poorly characterized; however, we shall see in Chapter 3 that when three or more spins are mutually coupled, structural information is available without a complete understanding of the "other" relaxation mechanisms. In any case, it is clear that relaxation mechanisms other than dipole–dipole are very important even though they do not explicitly contain structural information; this subject will be covered in Chapter 2.

A few interesting sidelights on Eq. (1.22) are now considered. If the ratio of the gyromagnetic ratios of the two spins is very large, one obviously will get the largest enhancements if the spin of lowest γ is observed while the spin of highest γ is saturated. Indeed, if the opposite experiment were done, the enhancement could easily be below the noise level. However, the advantage in observing the spin of lowest γ may be more than compensated for by the fact that the signal-to-noise problem is most severe for spins of low gyromagnetic ratio. If the saturated spin is an electron rather than a nucleus (the original Overhauser effect), the enhancements will be very large indeed. If protons which relax largely by dipole–dipole coupling to an electron are observed while the electron is saturated, enhancements on the order of -1000 will be observed. The "enhancement" is negative because the gyromagnetic ratio of the electron is negative, opposite in sign to that of the proton. These interesting effects, recently reviewed by Dwek *et al.* (9), are outside the scope of this work. It is interesting to note that several nuclei have negative magnetic moments so that a negative NOE would be expected for dipole–dipole relaxation; some examples are ^{15}N, ^{17}O, and ^{29}Si. A table of gyromagnetic ratios is given in Chapter 5.

G. Relaxation Times for Two Spins: The $\frac{3}{2}$ Effect

Despite the fact that T_1 is rigorously defined only for spins which obey Eq. (1.1), the literature is full of "T_1" values reported for spin systems which might reasonably be expected to obey Eq. (1.18). It is therefore appropriate to examine the meaning of T_1 when Eq. (1.18) applies; a more general and complete discussion of transient solutions of Eq. (1.18) will be given in Chapter 6.

1. Two Unlike Spins

Equations (1.18) form a set of coupled differential equations whose solutions depend on the boundary conditions used, i.e., the initial values of $\langle S_z \rangle$ and $\langle I_z \rangle$ at $t = 0$. If spin S is undisturbed while spin I is inverted by a 180° pulse or an adiabatic rapid passage, the initial conditions are

$$\langle I_z(t = 0) \rangle = -I_0, \qquad \langle S_z(t = 0) \rangle = +S_0$$

At a later time t we will have

$$[I_0 - \langle I_z \rangle]/I_0 = 2C \exp(-\lambda_1 t) + 2(1 - C) \exp(-\lambda_2 t) \qquad (1.26)$$

$$\left. \begin{matrix} \lambda_1 \\ \lambda_2 \end{matrix} \right\} = \tfrac{1}{2}(\rho_I + \rho_S) \pm \tfrac{1}{2}[(\rho_I - \rho_S)^2 + 4\sigma_{IS}\sigma_{SI}]^{1/2} \qquad (1.27)$$

$$C = (\lambda_1 - \rho_S)/(\lambda_1 - \lambda_2) \qquad (1.28)$$

Thus, in contradistinction to Eqs. (1.1) and (1.6), $\langle I_z \rangle$ is a double exponential function. Plots of $\log(I_0 - \langle I_z \rangle)$ versus t will not in general be linear.

Because of uncertainties in the initial values of $\langle I_z \rangle$ and the location of the origin, $t = 0$, the interpretation of the experimental data for a two spin system obeying Eq. (1.18) will generally involve a five-parameter least-squares analysis (cf. Chapter 6 and Appendix II, A). This procedure is not only complicated but requires very accurate data since so many parameters must be determined. However, in a great many cases, an experiment to measure "T_1" will give data which, on a $\log(I_0 - \langle I_z \rangle)$ versus time plot, is very nearly linear with a slope which can be related to a "T_1" by Eq. (1.6). In order to find the relationship between this experimental "T_1" and the relaxation parameters σ and ρ we shall examine two limiting cases of Eq. (1.26).

If $\rho_I \sim \rho_S \equiv \rho$ so that $\sigma \gg |\rho_I - \rho_S|$, Eq. (1.26) becomes

$$(I_0 - \langle I_z \rangle)/I_0 = 3 \exp[-(\rho - \sigma)t] - \exp[-(\rho + \sigma)t] \qquad (1.29)$$

But ρ is always at least twice σ and frequently larger, so Eq. (1.29) will be approximately exponential with

$$1/T_1 \simeq \rho_I$$

Next, consider the case when $\rho_I \neq \rho_S$ and $|\rho_I - \rho_S| \gg \sigma$. If we use the approximation

$$[1 + x]^{1/2} \simeq 1 + \tfrac{1}{2}x \qquad \text{with} \qquad x = \sigma^2/(\rho_I - \rho_S)^2$$

the exponents become

$$\left. \begin{matrix} \lambda_1 \\ \lambda_2 \end{matrix} \right\} = \rho_I \pm \frac{\sigma^2}{\rho_I - \rho_S} + \text{h.o.} \qquad (1.30)$$

where h.o. indicates higher-order terms in σ. Now

$$[I_0 - \langle I_z \rangle]/I_0 \simeq 2 \exp(-\rho_I t) \qquad (1.31)$$

and again an approximate single exponential is obtained.

We conclude: The answer to an ARP or pulsed NMR measurement of "T_1" when the decay is a single exponential is

$$1/T_1 \simeq \rho_I \qquad (1.32)$$

When there is no cross-relaxation, Eq. (1.32) is exact.

2. Identical Spins

A somewhat different approach is needed here. If I and S have the same Larmor frequency, then only $\langle I_z \rangle + \langle S_z \rangle$ can be observed. If I and S are not identical but are of the same nuclear species, one may still measure only $\langle I_z \rangle + \langle S_z \rangle$, either deliberately or because of the experimental technique used (e.g., pulsed NMR which always observes the sum of all nuclear spins of a given type). Adding the two equations (1.18a) and (1.18b) with $\rho_I = \rho_S$ (a likely case even when the spins are only nearly identical) one gets

$$d(\langle I_z \rangle + \langle S_z \rangle)/dt = -(\rho + \sigma)\,[\langle I_z \rangle + \langle S_z \rangle - I_0 - S_0] \qquad (1.33)$$

and

$$1/T_1 = \rho + \sigma \qquad (1.34)$$

Since, for dipole–dipole relaxation between I and S, $\sigma = \frac{1}{2}\rho$, we see that for dipole–dipole coupling of "identical" spins

$$1/T_1 = \tfrac{3}{2}\rho \qquad \text{(dipole–dipole only!)}$$

while for nonidentical spins $1/T_1 \simeq \rho$. This is the origin of the so-called "$\frac{3}{2}$ effect"† which has been a source of much confusion. It can now be seen that this "effect" is simply a matter of what is measured and, to some extent, how it is measured.

H. Conclusion

The close relationship between the nuclear Overhauser effect and nuclear spin–lattice relaxation has been established. Before considering how such experiments can be used to obtain structural or kinetic information about molecules, we must consider in more detail the mechanisms which cause spin–lattice relaxation.

---

† Abragam (7, p. 297) refers only to T_2 in this context, but subsequent authors have applied the term to T_1 as well.

REFERENCES

1. A. Carrington and A. D. McLachlan, "Introduction to Magnetic Resonance," Chapter 1. Harper, New York, 1967.
2. R. A. Hoffman and S. Forsén, *Progr. NMR Spectros.* **1**, 15 (1966).
3. R. Kaiser, *J. Chem. Phys.* **39**, 2435 (1963).
4. W. A. Anderson, *in* "NMR and EPR Spectroscopy" (Varian Staff). Pergamon, New York, 1960.
5. J. A. Pople, W. G. Schneider, and H. J. Bernstein, "High Resolution NMR." McGraw-Hill, New York, 1959.
6. J. H. Noggle, Ph.D. Thesis, Harvard Univ., Cambridge, Massachusetts, 1965.
7. A. Abragam, "The Principles of Nuclear Magnetism," Chapter VIII. Oxford Univ. Press, London and New York, 1961.
8. I. Solomon, *Phys. Rev.* **99**, 559 (1955).
9. R. A. Dwek, R. E. Richards, and D. Taylor, *Annu. Rev. NMR Spectros.* **2**, 293 (1969).

MECHANISMS OF SPIN–LATTICE RELAXATION

The most interesting mechanism of relaxation with respect to the nuclear Overhauser effect is the dipole–dipole mechanism. Except when chemical exchange processes are present (Chapter 7), dipole–dipole is the only common mechanism which contributes to σ as well as to ρ. However, in most cases, one or more other mechanisms may be effective and their contribution to ρ will reduce the observed NOE enhancements. These mechanisms must be taken into account before a quantitative interpretation of the NOE is possible.

When several mechanisms contribute to ρ, the total direct relaxation rate (henceforth called R) is

$$R = \sum_m \rho^m \tag{2.1}$$

where the summation is over all mechanisms m. Similarly, the transition probabilities between levels i and j must be summed over all possible mechanisms; the total transition probability is

$$W_{ij} = \sum_m W_{ij}^m \tag{2.2}$$

Spin–lattice relaxation is commonly discussed in terms of T_1; thus, when several mechanisms contribute,

$$1/T_1 = \sum_m 1/T_1^m \qquad (2.3)$$

For all mechanisms except dipole–dipole and scalar coupling

$$1/T_1^m = \rho^m$$

A. The Source of Spin–Lattice Relaxation

Every nucleus which has a spin also possesses a magnetic dipole moment μ; if a nuclear magnetic dipole is in a local magnetic field H_L, it will be coupled to that field with an energy

$$E = -\mu \cdot H_L \qquad (2.4)$$

If the molecule is in motion, as in a fluid state, then the local magnetic fields will fluctuate in time with, usually, an average value of H_L

$$\langle H_L(t) \rangle = 0$$

It is for this reason that dipole–dipole couplings, for example, are not observable in the spectrum of nuclei in liquids. However, if the fluctuations of H_L caused by the molecular motions have a component at the Larmor frequency ω_0 of the nuclear spins† this fluctuating field can induce transitions between the nuclear energy levels and cause relaxation. This process is similar to the effect of an rf magnetic field which oscillates at the Larmor frequency and induces the absorption which is observed as the NMR spectrum. These two phenomena are similar in that they both involve a magnetic dipole couple [Eq. (2.4)], but differ in that the rf absorption is caused by a coherent field at a single frequency whereas the fluctuations of H_L are random and have components at many frequencies. The result of the relaxation caused by the fluctuations of H_L is a transfer of energy between the spins and the molecular motions (the *lattice*) until thermal equilibrium is achieved. At thermal equilibrium the energy levels will have the populations required by the Boltzmann distribution, Eq. (1.8).

Typically, the translational and rotational motions of molecules in liquids will have components at the proper frequencies to cause spin–lattice relaxation. Vibrational motions, on the other hand, are usually too fast to be effective in

---

† More properly, the Fourier transform of the correlation function $\langle H_L(t) H_L(t+\tau) \rangle$ must have a nonzero component at ω_0.

relaxing spins. The effectiveness of internal rotation depends on the rate of rotation compared to the molecule's overall rotation rate; this subject will be discussed further in Chapter 4 and in Section D of this chapter.

The spin–lattice relaxation time is related to the mean-square average of the local magnetic field by (*1*)

$$1/T_1 = \tfrac{2}{3}\gamma^2 \langle H_{\mathrm{L}}\rangle^2 \tau_{\mathrm{c}}/(1 + \omega_0^2 \tau_{\mathrm{c}}^2) \tag{2.5}$$

where τ_{c} is the correlation time characterizing the motion which causes the relaxation. For small molecules in nonviscous liquids, the extreme narrowing condition,

$$\omega_0^2 \tau_{\mathrm{c}}^2 \ll 1$$

usually applies.

All nuclear interactions which contribute to H_{L} can cause relaxation. These various interactions are referred to as relaxation mechanisms; the more common mechanisms will be discussed individually in this chapter.

The preceding explanation of spin–lattice relaxation is not universally valid. Quadrupole relaxation (Section B), for example, is not a magnetic interaction in the sense of Eq. (2.4) but is, rather, an electrical interaction between the nuclear electric quadrupole moment and electric field gradients due to the surrounding electrons. However, the general features of Eq. (2.5) are found for all relaxation mechanisms. In particular, the relaxation time can be divided into two parts: the correlation time and a multiplier which we shall call the interaction constant. The interaction constant is a product of universal constants and molecular parameters such as quadrupole coupling constants or internuclear distances. It can often be calculated using data which are obtained by other experimental techniques. It is usually independent of temperature. On the other hand, the correlation time is difficult to calculate and independent experimental verification of its value is difficult and often impossible. It is the correlation time which is largely responsible for the temperature dependence of the spin–lattice relaxation time.

B. Quadrupole Relaxation

Any nucleus whose total spin quantum number $I > \tfrac{1}{2}$ has an electric quadrupole moment. This moment is usually denoted eQ; it is a property of the nucleus and is listed in most tables of nuclear properties. The interaction of nuclear quadrupole moments with electric field gradients provides a very efficient mechanism for coupling the nucleus to the rotational motions of the molecule. The electric field gradients are usually denoted eq and are caused by the electrons which surround the nucleus.

In the extreme narrowing limit, the quadrupole relaxation time is

$$1/T_1^Q = \rho^Q = K(I)\,(1 + \eta^2/3)\,(e^2\,qQ/h)^2\,\tau_c \qquad (2.6)$$

where q is the principal field gradient and η is the field gradient asymmetry. The effect of η can usually be neglected. The quantity e^2qQ/h is called the quadrupole coupling constant. The quadrupole coupling constant can be measured accurately for many compounds in the solid phase by nuclear quadrupole resonance (NQR) and in the gas phase by microwave spectroscopy. Values of e^2qQ/h have been determined for numerous nuclei and compounds and are compiled in many references (2a, b).

The constant K is

$$K(I) = (3\pi^2/10)\,(2I + 3)/I^2\,(2I - 1) \qquad (2.7)$$

and has numerical values

$$K(1) = 14.8044\,, \qquad K(\tfrac{3}{2}) = 3.94784$$

$$K(\tfrac{5}{2}) = 0.94748\,, \qquad K(\tfrac{7}{2}) = 0.40284$$

Electric field gradients are very small for free ions. In a compound, the more covalent the bond, the larger the field gradient. However, if a nucleus is in an electric field which has cubic (T_d, O_h, I_h) symmetry, the field gradients will again be very small regardless of how covalent the bonds may be. It should be noted that the electric field gradient is the *second* derivative of the electronic potential. If the nucleus is in a site of, for example, D_{3h} symmetry, the gradient of the electric potential is required to be zero at the site of the nucleus; the field gradient however is the second derivative of the potential and is not necessarily zero in such a site.

In a great majority of cases, quadrupole relaxation is the dominant relaxation mechanism of nuclei with $I > \tfrac{1}{2}$ and the relaxation times of such nuclei are very short. Because of rapid relaxation, scalar (J) couplings to such nuclei are usually not observed in the NMR spectrum of, e.g., protons. This does not mean that such coupling constants are zero; in fact, they may be quite large. What it does mean is that the rapid relaxation of the quadrupolar nucleus has effectively decoupled it from the protons in a manner similar (but *not* identical) to decoupling by NMR double resonance. The efficient relaxation of quadru-polar nuclei will cause their NOE enhancements to be very small and, probably, unmeasureable. Also, quadrupolar nuclei can usually be ignored when NOE experiments are performed on spin-$\tfrac{1}{2}$ nuclei.

Therefore, in general, quadrupolar nuclei relax efficiently and are not useful in NOE studies. Exceptions to this rule can occur: (a) If the nucleus is at a site of cubic symmetry. For example, the ^{14}N spectrum of $NH_4{}^+$ (aqueous, acid) is a sharp quintet; the ^{14}N spectrum of NH_3, on the other hand, is quite broad. (b) If the nucleus is ionic. As an extreme example, the relaxation time of Li^+ ion in D_2O (dilute solution) is $T_1 = 40$ sec (3). However, the relaxation time of aqueous Na^+ is much shorter, $T_1 = 0.057$ sec (4), presumably because its hydration sphere is less stable and/or less symmetrical. (c) If the nuclear electric quadrupole eQ is small. The most interesting example here is ^2D with $T_1 = 0.383$ sec for D_2O at 20°C, and $T_1 = 1.5$ sec in benzene-d_6 (5). Since proton relaxation times (cf. Section H) are usually one second or longer, such relaxation times are not prohibitively short for NOE studies. Even for deuterium, however, quadrupole relaxation seems to be the dominant mechanism in most molecules.

C. Intramolecular Dipole–Dipole Relaxation

Two magnetic dipoles at a distance r will couple with an energy

$$E_{dd} \propto \boldsymbol{\mu}_1 \cdot \boldsymbol{\mu}_2 / r^3$$

In liquids, this energy is averaged to zero by molecular rotations but can still cause nuclear relaxation. For two spins-$\frac{1}{2}$, I and S, separated by a distance r, the dipole–dipole (dd) mechanism contributes:

$$W_1{}^I(dd) = \frac{3}{20} \frac{\gamma_I{}^2 \gamma_S{}^2 \hbar^2}{r^6} \frac{\tau_c}{1 + \omega_I{}^2 \tau_c{}^2} \tag{2.8}$$

$$W_0(dd) = \frac{1}{10} \frac{\gamma_I{}^2 \gamma_S{}^2 \hbar^2}{r^6} \frac{\tau_c}{1 + (\omega_I - \omega_S)^2 \tau_c{}^2} \tag{2.9}$$

$$W_2(dd) = \frac{3}{5} \frac{\gamma_I{}^2 \gamma_S{}^2 \hbar^2}{r^6} \frac{\tau_c}{1 + (\omega_I + \omega_S)^2 \tau_c{}^2} \tag{2.10}$$

In most cases, extreme narrowing prevails and the frequency terms in the denominator can be neglected. In this limit the relaxation parameters defined by Eq. (1.18) are

$$\sigma_{IS}^{dd} = \tfrac{1}{2}\gamma_I{}^2 \gamma_S{}^2 \hbar^2 \tau_c / r^6 \tag{2.11a}$$

$$\rho_{IS}^{dd} = \gamma_I{}^2 \gamma_S{}^2 \hbar^2 \tau_c / r^6 \tag{2.11b}$$

If I and S are "identical" (cf. the "$\frac{3}{2}$ effect," Chapter 1, Section G)

$$1/T_1^{dd} = \tfrac{3}{2}\gamma^4 \hbar^2 \tau_c / r^6 \tag{2.12}$$

Cases where I and S are not spins-$\frac{1}{2}$ are of somewhat less interest. In most cases with S or I greater than $\frac{1}{2}$ quadrupole relaxation will dominate the ρ of that spin. In addition, most such nuclei have small gyromagnetic ratios so the dd contributions will be relatively small. In this more general case, the dipole–dipole relaxation parameters between spin I (total spin quantum number I) and spin S (quantum number S) are

$$\rho_{IS}^{dd} = \tfrac{4}{3}\gamma_I^2\,\gamma_S^2\,S(S+1)\,\hbar^2\,\tau_c/r^6 \tag{2.13a}$$

$$\rho_{SI}^{dd} = \tfrac{4}{3}\gamma_I^2\,\gamma_S^2\,I(I+1)\,\hbar^2\,\tau_c/r^6 \tag{2.13b}$$

$$\sigma_{IS}^{dd} = \tfrac{2}{3}\gamma_I^2\,\gamma_S^2\,I(I+1)\,\hbar^2\,\tau_c/r^6 \tag{2.14a}$$

$$\sigma_{SI}^{dd} = \tfrac{2}{3}\gamma_I^2\,\gamma_S^2\,S(S+1)\,\hbar^2\,\tau_c/r^6 \tag{2.14b}$$

Thus, the dipole–dipole interaction constant is related quite simply to molecular geometry. However, the correlation time is difficult to determine. While it is frequently convenient and reasonably accurate to assume simply that the correlation time is a molecular property and is therefore the same for all dipole–dipole interactions within a given molecule, it seems advisable at this point to give a detailed consideration to this complex and important property.

D. Rotational Correlation Times

The correlation time τ_c is the same for intramolecular dipole–dipole and quadrupole relaxation. It is most simply interpreted when the rotational motions of the molecule are in the rotational diffusion limit; i.e., the rotation proceeds by a small-step Brownian diffusion process with many steps required to reorient by one radian. Then the correlation time is related to the rotational diffusion constant D_r,

$$\tau_c = 1/6D_r \tag{2.15}$$

which is given in turn by

$$D_r = \langle \Delta\theta^2 \rangle / 2t \tag{2.16}$$

where $\langle \Delta\theta^2 \rangle$ is the net mean-squared angle turned in time t. Thus, in one correlation time the rms angle turned will be $3^{-\frac{1}{2}}$ radians ($\sim 33°$). For small molecules in nonviscous liquids, typically

$$10^{-10}\,\text{sec} > \tau_c > 10^{-12}\,\text{sec}$$

In molecules of less than cubic symmetry a single diffusion constant is not sufficient to characterize the rotational diffusion.† In the general case a rotational diffusion tensor is required (6). For rigid molecules of less than axial symmetry, the rotational diffusion tensor will have three independent components. Two independent components are needed for symmetric top molecules; one is needed for spherical tops. When internal motion is present, more than three diffusion constants may be required to totally characterize the molecule's rotational motion.

In many cases it is better to consider the reorientation of not the molecule, but the *relaxation vector*. For dipole–dipole relaxation, the relaxation vector is the vector connecting the two nuclei. For quadrupole relaxation, the appropriate vector is the principal axis of the field gradient tensor. If the quadrupolar nucleus is at the end of a bond, the principal axis is the bond axis; if it is centrally located, the principal axis is the symmetry axis. For example, for ^{11}B relaxation in BCl_3 the relaxation vector is the three-fold axis; for ^{14}N relaxation in CH_3CN it is the CN bond; for ^{35}Cl relaxation in CCl_3H it is the C–Cl bond. In unsymmetrical locations the relaxation vector for quadrupole relaxation may be difficult to locate (also η may be large).

In general one must assume that each relaxation vector has its own correlation time even though the number of possible independent parameters (diffusion constants) may be smaller than the number of correlation times. In CH_3CN and CD_3CN, even if one assumes no isotope effect, the H–H dipole–dipole correlation time is not the same as the deuterium quadrupole correlation time; the ^{14}N quadrupole correlation time is again different but related to the other two (see below). However, in large and unsymmetrical molecules it may be a good approximation to assume the correlation times are all the same. Exceptions to this rule may occur when rapid internal motion, as with methyl groups, is present.

The problem of anisotropic rotations and internal motions is complicated and largely unsolved. However, a few simple examples may help to clarify the subject. A rigid symmetric top molecule has two rotational diffusion constants: D_a the diffusion constant for rotation about the symmetry axis, and D_b the diffusion constant for rotation about any axis perpendicular to the symmetry axis. The correlation time of a relaxation vector which makes an angle θ to the principal axis of the symmetric top is (7, 8)

$$\tau_c = (3\cos^2\theta - 1)^2/24D_b + 3\sin^2\theta\cos^2\theta/(5D_b + D_a) + \tfrac{3}{4}\sin^4\theta/(2D_b + 4D_a)$$

$$(2.17)$$

For a spherical top ($D_a = D_b$), Eq. (2.17) reduces to Eq. (2.15), as expected, for any angle θ. Sizable differences have been found between D_a and D_b for

---

† This is similar to the situation found for the free rotation of a molecule where more than one moment of inertia is usually needed to characterize the molecules rotation.

small symmetric top molecules with dipole moments (9). Equation (2.17) presumes the rotational diffusion limit but the parallel motion (D_a) in particular has a tendency to be *inertial* (i.e., the rotational steps are large angle and the diffusion limit does not apply).

Some common cases can be illustrated with CH_3CN and CD_3CN. (We assume no isotope effect so that D_a and D_b are presumed to be the same for each of these molecules; this assumption is somewhat dubious especially for D_a.)

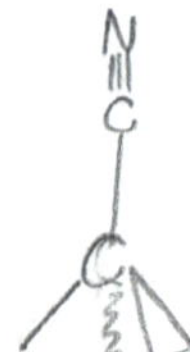

$$\theta = 0° \quad (^{14}N \text{ quadrupole})$$
$$\tau_c = 1/6D_b \tag{2.18a}$$

$$\theta = 90° \quad (H\text{–}H \text{ dipole–dipole})$$
$$\tau_c = \tfrac{1}{4}(6D_b)^{-1} + \tfrac{3}{4}(2D_b + 4D_a)^{-1} \tag{2.18b}$$

$$\theta = 109°\,28' \quad (^{2}D \text{ quadrupole, also } ^{13}C\text{–}H \text{ dipole–dipole})$$
$$\tau_c = \tfrac{1}{9}(6D_b)^{-1} + \tfrac{8}{27}(5D_b + D_a)^{-1} + \tfrac{16}{27}(2D_b + 4D_a)^{-1} \tag{2.18c}$$

Symmetric tops are, of course, not very common. However, a simple example can illustrate how the symmetric top formulas can be used in more complicated situations. Consider a methyl group which is attached to a large and more or less spherical molecule. We assume that the molecule as a whole reorients isotropically with a rotational diffusion constant D_0 and a correlation time $\tau_0 = 1/6D_0$. The methyl group has an internal rotation with a diffusion constant D_i and a correlation time τ_i which we shall define in terms of D_i as $\tau_i \equiv 1/2D_i$*. In Eq. (2.17) we use

$$D_a = D_0 + D_i \tag{2.19a}$$

$$D_b = D_0 \tag{2.19b}$$

For the proton–proton dd relaxation within the methyl group ($\theta = 90°$) we get

$$\tau_c = \tau_0[\tfrac{1}{4} + \tfrac{3}{4}(1 + 2\alpha)^{-1}] \tag{2.20a}$$

with $\alpha = \tau_0/\tau_i$. In the limits

---

* Defining τ_i with a two seems somewhat more appropriate than the six used for D_0 because D_i represents a one dimensional process. Also, such a definition makes τ_i closer to that used by Zeidler (10). In any case, it is only a definition and D_i is the fundamental quantity.

(1) Slow internal rotation, $\tau_i \gg \tau_0$,

$$\tau_c = \tau_0$$

(2) Fast internal rotation, $\tau_i \ll \tau_0$,

$$\tau_c = \tfrac{1}{4}\tau_0 + \tfrac{3}{8}\tau_i \sim \tfrac{1}{4}\tau_0 \tag{2.20b}$$

For ^{2}D relaxation in methyl-d_3 or ^{13}C–H dipole–dipole relaxation in methyl-^{13}C, ($\theta = 109°28'$) we get

$$\tau_c = \tau_0[\tfrac{1}{9} + \tfrac{3}{9}(1 + \alpha)/(1 + 2\alpha)(1 + \alpha/2)] \tag{2.21a}$$

In the limits

(1) Slow internal rotation; $\tau_c = \tau_0$, as before.
(2) Fast internal rotation;

$$\tau_c = \tfrac{1}{9}\tau_0 + \tfrac{8}{9}\tau_i \sim \tfrac{1}{9}\tau_0 \tag{2.21b}$$

It is surprising that the correlation time is different in the two limits but depends on τ_i in *neither* limit. Zeidler (*10*) obtained similar results which differ in detail from Eqs. (2.20) and (2.21); however, the difference is minor and his formulas yield the same limiting values of τ_c.

Wallach (*11*) has considered the case of a chain of repeated internal rotations by, essentially, applying Eq. (2.17) repeatedly to each rotation (as we did above for one rotation) and calculating its correlation time with respect to the coordinate frame of the preceding rotation. His formulas are particularly simple when the chain is attached to a large molecule (a protein is the example used) and all of the internal rotations are fast with respect to the rotation of the protein. In this limit only the first term of Eq. (2.17) is significant. According to Wallach, if τ_0 is the correlation time of the protein, the correlation time of a relaxation vector on a flexible chain is

$$\tau_c = \tau_0[\tfrac{1}{4}(3\cos^2\theta - 1)^2][\tfrac{1}{4}(3\cos^2\beta_{12} - 1)^2][\tfrac{1}{4}(3\cos^2\beta_{23} - 1)^2]\cdots \tag{2.22a}$$

where θ is the angle between the relaxation vector and the first internal rotation axis (on the outer end of the chain), β_{12} is the angle between the first and second internal rotation axes, etc. Equation (2.22a) contains a $\tfrac{1}{4}(3\cos^2\beta - 1)^2$ factor for each internal rotation in the chain which is faster than the rotation of the protein. It can be seen that for a single internal rotation in the fast limit, Wallach's formula is

$$\tau_c = \tau_0\tfrac{1}{4}(3\cos^2\theta - 1)^2 \tag{2.22b}$$

whose special cases [Eqs. (2.20b) and (2.21b)] were given previously.

These formulas have application, not only to NMR, but to ESR of spin labels and fluorescence polarization studies (*11*). In a later paper, Wallach (*12*) has presented experimental evidence and determined rates of internal rotation by measuring ^{14}N relaxation times in 2-, 3-, and 4-methylbenzylcyanide as a function of temperature.

All of the preceding formulas are placed in doubt if, as is likely, the extreme narrowing approximation fails for the protein or the rotation diffusion approximation fails for the internal rotation.

The correlation time is temperature dependent and is responsible for the temperature dependence of T_1 for quadrupole and dipole–dipole relaxation. The correlation time is shorter at higher temperatures, and its functional dependence on temperature is often found to be Arrhenius

$$\tau_c = \tau_c^{\circ} \exp(E_a/RT) \tag{2.23}$$

The activation energy is nearly that of the viscosity η. Typically, for un-associated, nonviscous liquids $\tau_c^{\circ} \sim 10^{-13}$ sec with E_a approximately 1–2 kcal mole^{-1}. For associated liquids, E_a is greater, e.g., 3.9 kcal in *n*-butyl alcohol (*13*).

The implication of Eq. (2.23) is that τ_c and therefore ρ^{dd} become larger at lower temperatures. If τ_c were to become sufficiently large, the extreme narrowing limit could be passed and ρ^{dd} would have a maximum, becoming smaller at yet lower temperatures. In practice, for small and unassociated molecules, the freezing point will be reached well before this and the extreme narrowing limit is the rule throughout the liquid range. The extreme narrowing limit is likely to be violated only for macromolecules or in highly associated and very viscous liquids.

Correlation times cannot be calculated with great accuracy, but the calculation using the microviscosity theory of Gierer and Wirtz (*14*) is sufficiently accurate for most purposes. The rotational diffusion constant can be related to the molecular friction constant β by

$$D = kT/\beta \tag{2.24}$$

Gierer and Wirtz calculated β by assuming a spherical solute molecule (radius r_0) surrounded by spherical solvent molecules (radius r_s) in a medium of viscosity η

$$\beta = 8\pi\eta r_0^3 f_r \tag{2.25}$$

with the microviscosity factor

$$f_r^{-1} = 6r_s/r_0 + 1/(1 + r_s/r_0)^3 \tag{2.26}$$

For pure liquids ($r_s = r_0$), $f_r = 1/6.125$.

Hence, the correlation time is

$$\tau_c = \eta V_m f_r / kT \tag{2.27}$$

$V_m = \frac{4}{3}\pi r_0^3$ is the volume of the solute molecule. The value V_m is usually estimated from the density of the solute by assuming that the pure solute is hexagonally close-packed with the molecules occupying 74% of the total volume; hence

$$V_m = 0.74 M_w / N_0 \rho \tag{2.28}$$

where N_0 is Avogadro's number, ρ is the density of the solute, and M_w is the gram molecular weight.

Van der Waals or covalent bond radii have also been used to estimate r_0. If the solute does not exist as a pure liquid at the desired temperature, Eq. (2.28) cannot be used and other methods must be found to estimate V_m; partial molal volumes are a possibility.

Equations (2.27) and (2.28) are useful for estimating dipole–dipole relaxation times, but one must be certain that the correlation time calculated by these equations is the one desired. These equations will work well for motions which reorient a dipole or for assymmetric molecules. Very symmetrical motions, particularly those which do not reorient a dipole and are not hindered by association, may proceed much more rapidly than expected from rotational diffusion theory or Eq. (2.27). These inertial effects will not only cause the T_1^{dd} calculation to be inaccurate but will make the spin–rotation mechanism important (see Section F).

To summarize: The correlation time for dipole–dipole and quadrupole relaxation will be longer (ρ larger, T_1 shorter) for large molecules, in viscous solutions, for associated molecules, and at low temperatures. Even though both σ and ρ are increased by these effects, it is probably best to avoid such conditions for NOE studies.

E. Intermolecular Dipole-Dipole Relaxation

Dipole–dipole interactions of nuclear spins on different molecules permit nuclear spin energy to be dissipated to the translational motions of the molecules. This mechanism is especially effective when high concentrations of nuclei with large magnetic moments (e.g., ^{1}H or ^{19}F) are present. It is largely a nuisance as far as NOE structural studies are concerned since it tends to decrease the intramolecular NOE enhancements. Kaiser (*15*), who observed an intermolecular NOE between the protons of cyclohexane and chloroform, suggested that such experiments should be of interest in studying the structure of liquids and intermolecular forces. A possible reason for the lack of such studies is the theoretical difficulties in treating both intermolecular relaxation and the theory of liquid mixtures. However, intermolecular NOE should be useful for studying association and solvation phenomena. In any case, we are primarily interested in intramolecular NOE and, therefore, in eliminating intermolecular effects.

The relaxation rate of spin I due to intermolecular dipole–dipole (xd) interactions with spin S is approximately

$$\rho_{IS}^{xd} = (8\pi/45)\, N_S\, \gamma_I^2\, \gamma_S^2\, \hbar^2\, S(S+1)/D_{IS}\, a \qquad (2.29)$$

where N_S is the concentration of spins S (spins per cubic centimeter), a is the distance of closest approach between spins I and S, and D_{IS} is the mutual (translational) self-diffusion constant of the molecules containing I and S.

$$D_{IS} = \tfrac{1}{2}(D_I + D_S) \qquad (2.30)$$

where D_I and D_S are the self-diffusion constants of I and S, respectively. These self-diffusion constants may be known from other data; if not, the diffusion constant can be estimated using the formula of Gierer and Wirtz (14) for the translational diffusion constant of a spherical molecule (radius r_0) in a spherical solvent (radius r_s) of viscosity η:

$$D(\text{trans}) = kT/\beta_t \qquad (2.31)$$

β_t is the translational friction constant,

$$\beta_t = 6\pi\eta r_0 f_t \qquad (2.32)$$

where f_t is the translational microviscosity factor

$$f_t^{-1} = (3r_s/2r_0) + 1/(1 + r_s/r_0) \qquad (2.33)$$

As with the rotational microviscosity theory, these formulas are commonly used for and probably work best for nonspherical molecules even though they were derived for spheres.

Equation (2.29) was derived under rather crude assumptions and, even if a good estimate of the diffusion constant is available, the poorly defined quantity a makes quantitative testing difficult. Nevertheless, the preceding equations indicate the major effects which are important in intermolecular dipole–dipole relaxation. The important factors are:

(a) ρ^{xd} is proportional to the concentration of the spins causing the relaxation;

(b) ρ^{xd} is proportional to $\gamma_S{}^2$ so spins of low gyromagnetic ratio are less effective;

(c) ρ^{xd} is smaller at higher temperatures and has a temperature dependence similar to that of the solvent's viscosity;

(d) ρ^{xd} will be larger in more viscous solvents.

A good discussion of alternative theoretical approaches is given by Hertz (*16*). A test of how well Eqs. (2.29)–(2.33) work, as well an idea of the magnitudes of xd relaxation effects, is given by Prendred *et al.* (*17*) and Hertz (*16*).

For xd, as for dd relaxation,

$$\sigma_{IS}^{xd} = [I(I + 1)/2S(S + 1)]\,\rho_{IS}^{xd} \tag{2.34}$$

ρ_{SI}^{xd} and σ_{SI}^{xd} can be obtained by exchanging I and S in Eqs. (2.29) and (2.34). Again, $1/T_1$ for "like" spins is $\frac{3}{2}\rho$ and $1/T_1$ for "unlike" spins is approximately ρ.

In an NOE experiment, four sources of intermolecular dipole–dipole relaxation can be distinguished: identical solute molecules, solvent, lock sample (if internal lock is used), and paramagnetic impurities.

For relaxation due to identical solute molecules, only the N_S factor of Eq. (2.29) is under the experimenter's control as an effective means to reduce ρ^{xd}. The sample concentration must be kept low so that the intermolecular contributions to ρ and σ are small compared to the intramolecular effects. Only under these conditions will NOE enhancements have a direct and simple relationship to molecular structure.

The other sources of xd relaxation will reduce the observed NOE and must therefore be minimized. Use of nonmagnetic or low magnetic solvents such as CS_2, OCS, or CCl_4 is recommended. Deuterated solvents are less desirable but, since they are now widely available in high purity, they are likely to be a common choice. The relaxation effect of deuterium (spin 1) is reduced from that of protons at the same concentration by

$$\frac{8}{3}(\gamma_D{}^2/\gamma_H{}^2) = 0.06 \tag{2.35}$$

While this effect is not entirely negligible, it is sufficiently small for most purposes. The effects of solvent and internal lock sample will be discussed further in Chapter 5.

Because of the large magnetic moment of an unpaired electron, small amounts of paramagnetic impurities can give a sizable contribution to the xd spin–lattice relaxation rate. All such impurities, including dissolved oxygen, must be rigorously eliminated. As an example, Gutowsky and Woessner (*18*) found the proton and fluorine T_1's in 1,3,5-trifluorobenzene to be reduced from 26 to 1.8 sec and 17 to 1.8 sec, respectively, if dissolved oxygen was not removed. Effects of this magnitude could totally obscure an NOE.

F. Spin–Rotation Relaxation

The interaction of the nuclear magnetic moment and the rotational magnetic moment of the molecule in which the nucleus is situated provides a direct mechanism for the transfer of nuclear spin energy to the molecular rotation. This mechanism is distinctly different from the indirect transfer which occurs by the dipole–dipole or quadrupole mechanisms. If $\mathbf{I}$ is the spin angular momentum vector and $\mathbf{J}$ is the molecular rotation angular momentum vector, the spin rotation coupling energy is of the form

$$E_{SR} \propto \mathbf{I} \cdot \mathbf{C} \cdot \mathbf{J}$$

where $\mathbf{C}$ is the spin–rotation coupling tensor. This coupling is very weak and its effect can be observed only by molecular beam magnetic resonance or very high resolution rotational (microwave) spectroscopy.

The fluctuations of the spin–rotation energy form an important mechanism of relaxation, particularly for spin-$\frac{1}{2}$ nuclei where the quadrupole mechanism is not available. In the gas phase it is frequently the only effective relaxation mechanism for spin-$\frac{1}{2}$ nuclei. Spin–rotation can be an effective mechanism in liquids in circumstances when "free" or "gaslike" rotations are present; this is the inertial limit, the opposite of the diffusion limit discussed previously. Inertial effects and spin–rotation relaxation will be important at high temperatures, for symmetrical rotations, and for molecules without dipole moments; they are frequently associated with anomalously low viscosities. Inertial and diffusion effects can be important simultaneously for different motions in the same molecule.

According to Hubbard (*19*),† (*20*)‡ the spin–rotation (SR) relaxation rate is

$$\rho^{SR} = 1/T_1^{SR} = (2IkT/3\hbar^2)\, C^2\, \tau_{SR} \tag{2.36}$$

where I is the moment of inertia, C^2 is the squared average of the SR tensor, and τ_{SR} is the spin–rotation correlation time. While Eq. (2.36) is only valid for spherical molecules in the extreme narrowing limit when $\tau_c \gg \tau_{SR}$, it is sufficient to show the major effects. So little is known about spin–rotation that more sophisticated versions are hardly worthwhile; for example, τ_{SR} is poorly

---

† Note that Hubbard's use of Stoke's law makes many of his equations incorrect.

‡ The derivation of Green and Powles (*20*) is less convincing but more understandable than that of Reference *19*.

understood and C^2 has been measured for only a few simple molecules, mostly diatomics. However, C^2 can sometimes be estimated from chemical shift data (*21*).

The temperature dependence of spin–rotation relaxation is largely due to τ_{SR}. Experimentally τ_{SR} and ρ^{SR} are found to *increase* at *higher* temperatures; this is the opposite of the behavior found for dipole–dipole relaxation. It is not uncommon for both dd and SR relaxation to be effective for spin-$\frac{1}{2}$ nuclei in the liquid phase. When this occurs, dd will dominate at *low* temperatures and SR at *high* temperatures and ρ will have a minimum value at some intermediate temperature (T_1 will have a maximum). Such behavior is illustrated in Fig. 2-1 for the relaxation of ^{13}C in methyl iodide (*22*).

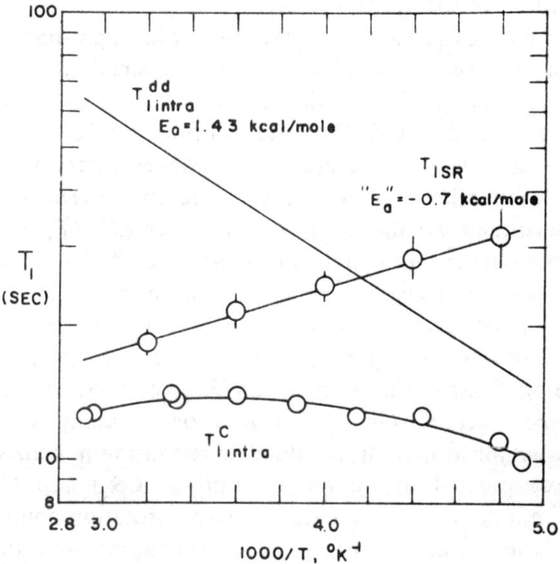

Fig. 2-1 Carbon-13 relaxation times in methyl iodide–^{13}C. T_1^{dd}(intra) is calculated from other data and subtracted from T_1^{-1}(experimental) to separate the dipole–dipole and spin–rotation contributions. Reproduced with permission from Gillen *et al.* (*22*).

The temperature dependence of τ_{SR} was calculated for spherical molecules by Hubbard (*19*) who derived the relationship

$$\tau_c \, \tau_{SR} = I/6kT \tag{2.37}$$

I is again the moment of inertia. Equation (2.37) combined with Eqs. (2.36) and (2.11) predict

$$T_1^{SR} \, T_1^{dd} = \text{constant} \tag{2.38}$$

These equations predict that the SR temperature dependence is equal and opposite to that of dd [Eqs. (2.23) and (2.27)] and provide a method for calculating τ_{SR}. However, they are difficult to confirm experimentally because so little is known about SR coupling constants. Equations (2.37) and (2.38) must be used with caution when anisotropic rotation is likely (*22*).

Spin–rotation relaxation has been found to be important for ^{19}F relaxation in symmetrical molecules: SF_6 (*23*) ($T_1 = 0.5$ sec),[†] MoF_6 (*24*) ($T_1 = 0.85$ sec) and CF_4 (*25*) ($T_1 = 3.1$ sec at 143°K, T_1 has a maximum value of 3.6 sec at 118°K). It is likewise important for ^{31}P relaxation in P_4O_6 (*26*) ($T_1 \sim 11$ sec). It is also important in less symmetrical small molecules: for ^{31}P in PCl_3 ($T_1 \sim 1.5$ sec) and $POCl_3$ ($T_1 \sim 5$ sec) (*27*) and ^{19}F in AsF_3 ($T_1 \sim 3$ sec with a maximum near room temperature[‡]) (*28*).

Spin–rotation appears to be less important for protons, presumably because the proton spin–rotation coupling constants are smaller. However, SR relaxation has been found in acetonitrile (*29*) (maximum $T_1 \sim 22$ sec at room temperature when diluted in CD_3CN) and the proton relaxation of methyl iodide (*22*) (maximum $T_1 \sim 15$ sec near room temperature when diluted in CD_3I). Spin–rotation relaxation was also found in benzene (*30*) at high temperatures (maximum T_1 above 200°C), in water (*31*) (T_1 maximum at 180°C) and in ammonia (*31*) (maximum $T_1 \sim 50$ sec at 0°C).

It may seem from the preceding that spin–rotation relaxation is limited to small or highly symmetrical molecules. This is true in large part; the factors mentioned previously as causing τ_c to become long will also prevent spin–rotation from being effective. Burke and Chan (*32*), however, have emphasized the potential importance of spin–internal-rotation coupling as a source of relaxation. They demonstrate that, for fluorine relaxation in benzotrifluoride ($T_1 = 3.0$ sec extrapolated to infinite dilution in CS_2) and CF_3CCCF_3 ($T_1 = 2.8$ sec at infinite dilution in CCl_4), the spin–rotation coupling to the internal CF_3 rotation is an important relaxation mechanism. In benzotrifluoride this mechanism dominates at room temperature and below.

For spin–internal-rotation relaxation, ρ is still larger at higher temperatures, but the temperature dependence is not as strong. This is presumably because, in these molecules at least, the barrier to internal rotation is smaller than for the overall molecular rotation.

Burke and Chan also reexamined the work of Pritchard and Richards (*33*) on the relaxation of the methyl protons of toluene and *p*-xylene (CS_2 solvent) and concluded that intermolecular dipole–dipole relaxation accounts for only one-third of the observed relaxation rate (i.e., one-third of R or $1/T_1$ at infinite

---

[†] Unless otherwise specified, relaxation times given are at or near room temperature (25±5°C).

[‡] The competing mechanism is not, however, dipole–dipole but scalar coupling (cf. Section G).

dilution in CS_2). They suggest that spin–rotation interactions with the methyl internal rotation accounts for the remainder and is therefore the dominant relaxation mechanism at room temperature. The facile reorientation of methyl groups is probably also responsible for the importance of spin–rotation relaxation in acetonitrile (*29*) and methyl iodide (*22*)—it is not properly termed an internal rotation in these cases but the principle is the same.

In conclusion: Freely rotating groups such as CH_3 and CF_3 may relax by spin–rotation even when they are part of a large molecule which would not ordinarily be suspected of spin–rotation interactions. Spin–rotation may also be found in symmetrical molecules in nonviscous and nonpolar solvents, especially at high temperatures.

G. Relaxation by Scalar Coupling

By scalar coupling we mean simply the "*J* coupling" responsible for the familiar multiplet splittings observed in high resolution NMR. It is of the form

$$E_{SC} = J\mathbf{I} \cdot \mathbf{S}$$

This coupling can become a source of relaxation for spin I if either J or S becomes time dependent. J can become time dependent as a result of chemical exchange or internal rotation; discussion of this important subject will be deferred until Chapter 7. Here we shall discuss relaxation by scalar coupling modulated by rapid relaxation of spin S; i.e., S is time dependent. Under these conditions, of course, the multiplet splitting of S on I will not be observed since $\langle E_{SC} \rangle \cong 0$.

According to Abragam (*34*), for relaxation due to scalar coupling of spin I modulated by relaxation of spin S

$$\rho_{IS}^{SC} = \frac{1}{T_1^{SC}} = \frac{[(\tfrac{8}{3})\pi^2 J^2 S(S+1)]\,T_{2S}}{1 + (\omega_I - \omega_S)^2\,T_{2S}^2} \tag{2.39}$$

$$\sigma_{IS}^{SC} = -\,[I(I+1)/2S(S+1)]\,\rho_{IS}^{SC} \tag{2.40}$$

$$\frac{1}{T_{2I}^{SC}} = [\tfrac{4}{3}\pi^2 J^2 S(S+1)]\left[T_{2S} + \frac{T_{2S}}{1 + (\omega_I - \omega_S)^2\,T_{2S}^2}\right] \tag{2.41}$$

From Eq. (2.39) it can be seen that unless T_{2S} is so short that the linewidth of S is of the same magnitude as the difference of the Larmor frequencies ρ^{SC} will be very small. Therefore a combination of very rapid S relaxation and nearly equal gyromagnetic ratios is required for scalar coupling to be important. Rhodes *et al.* (*35*) found scalar coupling of ^{31}P to ^{81}Br and ^{79}Br to be significant in the relaxation of ^{31}P in PBr$_3$ at 9 and 18 MHz. Spin–rotation relaxation is also important for ^{31}P. The relevant quantities are

$$\omega_0(^{31}\text{P}) = 2\pi \cdot (18\,\text{MHz})$$

$$\omega_0(^{81}\text{Br}) = 2\pi \cdot (12\,\text{MHz})$$

$$T_2(^{81}\text{Br}) = 3.3 \times 10^{-7}\,\text{sec}$$

$$J(^{81}\text{Br} - {}^{31}\text{P}) = 370\,\text{Hz}$$

On the other hand, scalar coupling to ^{35}Cl and ^{37}Cl does not contribute significantly to the ^{31}P relaxation in PCl$_3$ and POCl$_3$ at 18 MHz (*27*). The Larmor frequency of ^{35}Cl, $2\pi \cdot (4.35\,\text{MHz})$, is much smaller than that of ^{81}Br and its T_2 will be one or two orders of magnitude larger. In this case therefore, the frequency term in the denominator of Eq. (2.39) is large compared to one and ρ^{SC} will be small.

Because of Eq. (2.40), which predicts a negative NOE if scalar coupling dominates the spin–lattice relaxation, the possibility of proton–proton scalar coupling relaxation has been of some concern. For like nuclei we define the chemical shift (in hertz) between I and S as

$$\delta = (\omega_{\text{I}} - \omega_{\text{S}})/2\pi \qquad (2.42\text{a})$$

and the linewidth of spin S (not including contributions from the magnetic field inhomogeneity) as

$$\Delta_{\text{S}} = 1/\pi T_{2\text{S}} \qquad (2.42\text{b})$$

For spins-$\frac{1}{2}$, Eq. (2.39) becomes

$$\rho^{\text{SC}} = 2\pi(J^2/\Delta_{\text{S}})/(1 + 4\delta^2/\Delta_{\text{S}}^2) \qquad (2.43)$$

If $\Delta_{\text{S}} \gtrsim \delta$, NOE experiments between I and S will be impossible because their resonances will overlap.

If $\Delta_{\text{S}} \ll \delta$, Eq. (2.43) becomes

$$\rho^{\text{SC}} = (\pi/2)\,\Delta_{\text{S}}(J/\delta)^2 = (1/2T_{2\text{S}})\,(J/\delta)^2 \qquad (2.44)$$

However, $\Delta_{\text{S}} \ll \delta$ implies that $(1/2T_{2\text{S}})$ will be quite small. The effect is further reduced by the J/δ factor which must be much less than one for loosely coupled spins and ρ^{SC} will usually be insignificant compared to ρ^{dd}. If I and S are tightly coupled, the theory of the NOE presented here is inadequate (cf.

Appendix I); the more exact density matrix relaxation theory, needed for tightly coupled spins, automatically accounts for such effects.

It is therefore unlikely that scalar coupling relaxation in the absence of chemical exchange effects will be a significant mechanism of relaxation in any nuclear spin system for which the NOE is apt to be feasible or interesting. On the other hand, NOE experiments on systems with scalar coupling modulated by chemical exchange not only can be but have been done; these will be discussed in Chapter 7.

Scalar coupling modulated by relaxation more often contributes to T_2 (i.e., the linewidth) through the frequency independent term of Eq. (2.41). A familiar example of such an effect is the broad lines commonly observed for protons coupled to nitrogen, especially those directly bonded to nitrogen. Such broadenings do not necessarily mean that ρ^{SC} is significant or that NOE measurements involving such protons will be in any way affected. Indeed, N–H couplings are quite unlikely to affect proton spin–lattice relaxation rates at all and their only effect on NOE experiments will be that more power may be needed to saturate the resonances which are broadened by N–H scalar coupling. Of course, if the N–H protons are labile, chemical exchange effects must be considered.

H. Relaxation Times of Protons

In Tables 2-1–2-3 of this section are listed typical proton relaxation times for a number of small molecules. The "intramolecular" contributions to T_1^{-1} were measured by extrapolating the relaxation times of the protons in the

TABLE 2-1

RELAXATION TIMES OF AROMATIC PROTONS

Molecule	T_1 (neat) (sec)	T_1 (intra) (sec)	T_1 (inter) (sec)	Solvent	Reference
Benzene	22	100	29	CS_2	*34*
Benzene	19	110	23	deut[a]	*10, 20*
Pyridene	12	100	14	deut	*10, 20*
Toluene (ring)	16	56	23	CS_2	*34*
p-Xylene (ring)	14	45	21	CS_2	*10, 20*
o-Xylene (ring)	12	56	16	CS_2	*10, 20*
Mesitylene	10	40	13	CS_2	*38, 20*
Chlorobenzene	18	80	23	CS_2	*39*
Bromobenzene	14	67	18	CS_2	*39*

[a] The abbreviation "deut" means the deuterated analog of the compound at the left.

compound being investigated to infinite dilution. The dilution is carried out either in the deuterated analog of the compound being investigated or in a solvent such as CS_2 or CCl_4. The former procedure is clearly the best since it will yield a true intramolecular T_1; however, the intermolecular D–H dipole–dipole coupling at infinite dilution must be calculated and subtracted from the extrapolated T_1^{-1} to yield the true T_1 (intra). Extrapolations in CCl_4 or CS_2

TABLE 2-2

RELAXATION TIMES OF METHYL PROTONS

Molecule	T_1 (neat) (sec)	T_1 (intra) (sec)	T_1 (inter) (sec)	Solvent	Reference
CH_3Br	~5	~5	—	CS_2	40
CH_3I	14	20	45	CS_2	40[a]
CH_3I	12.5	16	62	deut[b]	10
CH_3CN	15	22	45	deut	30
Acetone	16	29	34	deut	10
$(CH_3)_2SO$ (DMSO)	2.9	5.6	6.1	deut	10
Toluene (methyl)	10	~15	—	CS_2	34
p-Xylene (methyl)	7.4	10.5	~30	CS_2	34
o-Xylene (methyl)	6.7	~12	—	CS_2	34
Acetic Acid-d_1	4.8	10.2	9.1	deut	10

[a] For methyl iodide see also Gillen *et al.* (*22*).

[b] The abbreviation "deut" means the deuterated analog of the compound at the left.

TABLE 2-3

OTHER PROTON RELAXATION TIMES

Molecule	T_1 (neat) (sec)	T_1 (intra) (sec)	T_1 (inter) (sec)	Solvent	Reference
CH_2Cl_2	30	67	—	CS_2	17
CH_2Br_2	16	42	—	CS_2	17
CH_2I_2	6.6	30	7.8	CS_2	17
$CHCl_3$	84	333	117	CS_2	17
$CHBr_3$	22	154	26	CS_2	17
H_2O	3.6	—	a	—	41
n-C_5H_{10}	14.3	11.5	b	CCl_4	42
n-C_8H_{18}	5.5	5.0	b	CCl_4	42
n-$C_{12}H_{26}$	2.0	2.5	b	CCl_4	42
n-$C_{16}H_{34}$	1.0	1.8	b	CCl_4	42

[a] T_1 (inter) is calculated as 10 sec but cannot be measured in the usual way.

[b] The concentration dependence of T_1 is primarily a viscosity effect so a meaningful T_1 (inter) cannot be calculated from this data.

assume that these compounds have intermolecular interactions similar to those of the compound being investigated. This assumption is often reasonable, but is clearly invalid for associated liquids. The intermolecular T_1 is then calculated from

$$T_1^{-1}(\text{inter}) = T_1^{-1}(\text{neat}) - T_1^{-1}(\text{intra}) \qquad (2.45)$$

For protons, the intramolecular relaxation may contain contributions from spin–rotation as well as dipole–dipole relaxation

$$T_1^{-1}(\text{intra}) = (T_1^{SR})^{-1} + (T_1^{dd})^{-1} \qquad (2.46)$$

It should be noted that many of the references neglected the possibility of a spin–rotation contribution and their analyses of the data, particularly the calculations of τ_c, are accordingly suspect. Spin–rotation relaxation is most likely for methyl groups.

Many of the relaxation times on the tables have been read from graphs in the literature and may not be exact. Likewise, the T_1 (intra) is often not corrected for the residual D–H intermolecular relaxation in experiments where extrapolations were carried out in deuterated solvents (indicated deut on the tables). For exact values, the reader should consult the original references.

Other compilations of relaxation times are those of Hertz (16) and Prendred et al. (17).

I. Conclusion

The most important relaxation mechanisms for spin-$\frac{1}{2}$ nuclei in the liquid phase are intramolecular dipole–dipole, intermolecular dipole–dipole, and spin–rotation. Relaxation by scalar coupling is unlikely to be important for systems of interest with respect to the NOE unless chemical exchange occurs. The frequently discussed but seldom found mechanism, anisotropic chemical shift relaxation (36),† appears to be rare. Perhaps the only definite and unambiguous example of this mechanism is in $CHFCl_2$ at very low temperatures (37).‡ However, since for the anisotropic chemical shift mechanism T_1^{-1} is proportional to H_0^2, it may be found more frequently with modern spectrometers which operate at higher fields—especially for ^{13}C (43).

---

† Note that examples cited by Becker (36) are probably spin–rotation rather than anisotropic shielding relaxation.

‡ The relaxation times were measured at $-142°C$, ^{19}F NMR at 56.4 MHz, 30% (v/v) in OCS. Even here, spin–rotation relaxation dominates $T_1(F)$ at higher temperatures, cf. Chapter 8, Section D.

We shall call the dipole–dipole relaxation between spins i and j, ρ_{ij} and σ_{ij}. In these terms, the total relaxation rate of spin i is

$$R_i = \sum_{j \neq i} \rho_{ij} + \rho_i^{\mathrm{xd}} + \rho_i^{\mathrm{SR}} \tag{2.47}$$

where the sum is taken over all of the other nuclei in the molecule and, unless low concentrations are used, may contain inter and intra contributions. Therefore, ρ^{xd} includes only solvent, lock sample, and impurity contributions. The last two terms of Eq. (2.47) will frequently be characterized as "other" relaxation mechanisms and be called

$$\rho_i^* = \rho_i^{\mathrm{xd}} + \rho_i^{\mathrm{SR}} \tag{2.48}$$

Other relaxation mechanisms such as quadrupole or anisotropic chemical shift may be included in ρ^* when appropriate.

REFERENCES

1. C. P. Slichter, "Principles of Magnetic Resonance," p. 153. Harper, New York, 1963.
2a. S. L. Segal and R. G. Barnes, Catalogue of nuclear quadrupole interactions and resonance frequencies in solids. U.S.A.E.C. Res. and Develop. Rep. Pt. I (IS-520, 1968) and Pt. II (IS-1222, 1965). Available from Clearing house for Fed. Sci. and Tech. Inform., Nat. Bur. of Stand., Springfield, Virginia.
2b. B. Starck, Molecular constants from microwave spectroscopy. *In* "Numerical Data and Functional Relationships in Science and Technology" (K.-H. Hellwege and A. M. Hellwege, eds.), Group II, Vol. 4. Springer, Berlin, 1967.
3. D. E. Woessner, B. S. Snowden, Jr., and A. G. Ostroff, *J. Chem. Phys.* **49**, 371 (1968).
4. T. L. James and J. H. Noggle, *J. Amer. Chem. Soc.* **91**, 3424 (1969).
5. D. E. Woessner, *J. Chem. Phys.* **40**, 2341 (1964).
6. W. T. Huntress, Jr., *Advan. Magn. Resonance* **4**, 2 (1970).
7. H. Shimizu, *J. Chem. Phys.* **40**, 754 (1964); **37**, 765 (1962).
8. D. E. Woessner, *J. Chem. Phys.* **37**, 647 (1962).
9. K. T. Gillen and J. H. Noggle, *J. Chem. Phys.* **53**, 801 (1970).
10. M. D. Zeidler, *Ber. Bunsenges. Phys. Chem.* **69**, 659 (1965).
11. D. Wallach, *J. Chem. Phys.* **47**, 5258 (1967).
12. D. Wallach, *J. Phys. Chem.* **73**, 307 (1969).
13. R. A. Bernheim, H. S. Gutowsky, and I. J. Lawrenson, *J. Chem. Phys.* **34**, 565 (1961).
14. A. Gierer and K. Wirtz, *Z. Naturforsch. A* **8**, 532 (1953).
15. R. Kaiser, *J. Chem. Phys.* **42**, 1838 (1965).
16. H. G. Hertz, *Progr. NMR Spectros.* **3**, 159 (1967).
17. T. L. Prendred, A. M. Pritchard, and R. E. Richards, *J. Chem. Soc. A* p. 1009 (1966).
18. H. S. Gutowsky and D. E. Woessner, *Phys. Rev.* **104**, 843 (1956).
19. P. S. Hubbard, *Phys. Rev.* **131**, 1155 (1963).
20. D. K. Green and J. G. Powles, *Proc. Phys. Soc. London* **85**, 87 (1965).
21. C. Deverell, *Mol. Phys.* **18**, 319 (1970); **17**, 551 (1969); and references cited therein.
22. K. T. Gillen, M. Schwartz, and J. H. Noggle, *Mol. Phys.* **20**, 899 (1970).

23. W. S. Hackleman and P. S. Hubbard, *J. Chem. Phys.* **39**, 2688 (1963).
24. P. Rigney and J. Virlet, *J. Chem. Phys.* **47**, 4645 (1967).
25. J. S. Rugheimer and P. S. Hubbard, *J. Chem. Phys.* **39**, 552 (1963).
26. D. J. Mowthorpe and A. C. Chapman, *Mol. Phys.* **15**, 429 (1968).
27. D. W. Aksnes, M. Rhodes, and J. G. Powles, *Mol. Phys.* **14**, 333 (1968).
28. M. Rhodes and D. W. Aksnes, *Mol. Phys.* **17**, 261 (1969).
29. D. E. Woessner, B. S. Snowden, Jr., and E. T. Strom, *Mol. Phys.* **14**, 265 (1968).
30. J. G. Powles and R. Figgins, *Mol. Phys.* **10**, 155 (1966).
31. D. W. G. Smith and J. G. Powles, *Mol. Phys.* **10**, 451 (1966).
32. T. E. Burke and S. I. Chan, *J. Magn. Resonance* **2**, 120 (1970).
33. A. M. Pritchard and R. E. Richards, *Trans. Faraday Soc.* **62**, 1388 (1966); also Reference *20*.
34. A. Abragam, "The Principles of Nuclear Magnetism," p. 311. Oxford Univ. Press, London and New York, 1961.
35. M. Rhodes, D. W. Aksnes, and J. H. Strange, *Mol. Phys.* **15**, 541 (1968).
36. E. D. Becker, "High Resolution NMR," p. 205. Academic Press, New York, 1969.
37. E. L. Mackor and C. MacLean, *Progr. NMR Spectros.* **3**, 152 (1967).
38. C. A. Reilly and R. L. Strombothe, *J. Chem. Phys.* **26**, 1338 (1957).
39. R. W. Mitchell and M. Eisner, *J. Chem. Phys.* **33**, 86 (1960).
40. A. M. Pritchard and R. E. Richards, *Trans. Faraday Soc.* **62**, 2014 (1966).
41. K. Krynicki, *Physica (Utrecht)* **32**, 167 (1967).
42. D. E. Woessner, *J. Chem. Phys.* **41**, 84 (1964).
43. H. W. Spiess, D. Schweitzer, U. Haeberlen, and K. H. Hausser, paper presented at the 12th Experimental NMR Conference, Gainesville, Florida, February, 1971.

CHAPTER

3

THE NUCLEAR OVERHAUSER EFFECT IN RIGID MOLECULES

The results of Chapter 2 indicate that nuclear relaxation times by themselves do not furnish much information on molecular geometry. Only intramolecular dipole–dipole relaxation depends explicitly on internuclear distances and, even if the relaxation effect of this mechanism can be separated from the other contributing mechanisms, it is difficult to determine the correlation times independently. In this chapter we shall consider steady-state NOE experiments on rigid molecules in the absence of chemical exchange phenomena; molecules with internal motion and applications of NOE to chemical exchange will be covered in later chapters. We shall also defer until later the subject of transient NOE. Transient experiments can provide valuable supplementary information to steady-state NOE but in this chapter we shall show how the exclusive use of steady-state results can give information on molecular geometry.

A. Relaxation in Multispin Systems

Without delving deeply into the density matrix theory of nuclear relaxation or the derivation of the two spin–relaxation equations from this theory, Eqs. (1.18) can be generalized quite simply by considering a multispin system as a sum of pairwise interactions and summing the relaxation effects of each

44

pair of spins. For a spin i coupled to a group of spins j, we have

$$d\langle \mathbf{I}_{zi}\rangle/dt = -R_i(\langle \mathbf{I}_{zi}\rangle - I_{0i}) - \sum_{j\neq i} \sigma_{ij}(\langle \mathbf{I}_{zj}\rangle - I_{0j}) \qquad (3.1)$$

$$R_i = \sum_{j\neq i} \rho_{ij} + \rho_i^* \quad \Leftarrow ok \ for \ all \ cases \ of \ \tau c \qquad (3.2)$$

The quantities in this equation have simple experimental interpretations:

(a) $\langle \mathbf{I}_{zi}\rangle$ is proportional to the total integrated intensity of the NMR resonance of spin i measured under slow-passage, nonsaturating, conditions.
(b) I_{0i} is the value of $\langle \mathbf{I}_{zi}\rangle$ at thermal equilibrium.
(c) R_i is the total direct relaxation rate of spin i; a sum of ρ for spin i over all possible relaxation paths. Except when i is one of a group of equivalent spins, the relaxation time of spin i is roughly $T_{1i} \sim R_i^{-1}$. The meaning of T_1 for groups of equivalent spins will be discussed in Section D (p. 51).
(d) σ_{ij} is the cross-relaxation between spins i and j. In this chapter we shall consider cases where only dipole–dipole relaxation contributes to σ.
(e) ρ_{ij} is the direct dipole–dipole relaxation between spins i and j. If the sample whose structure is being studied is at sufficient dilution only *intra*-molecular dipole–dipole interactions of spins i and j will contribute to ρ_{ij} and σ_{ij}. In addition, we shall be primarily interested in cases when all spins are spin-$\frac{1}{2}$. Under these conditions

$$\sigma_{ij} = \sigma_{ji} \qquad (3.3a)$$

$$\rho_{ij} = \rho_{ji} = 2\sigma_{ij} \quad \Leftarrow extreme \ narrowing \ condition \ only \qquad (3.3b)$$

$$\rho_{ij} = \gamma_i^2 \gamma_j^2 \hbar^2 \tau_c(ij)/r_{ij}^6 \quad \Leftarrow \qquad (3.4)$$

see p 25

$\tau_c(ij)$ is the correlation time of the vector between spins i and j.
(f) ρ_i^* is the direct relaxation of spin i due to "other" relaxation paths. It includes:

    (*i*) dipole–dipole interactions with spins not included in the sum over j [Eq. (3.2)] (e.g., solvent and lock-sample spins);
    (*ii*) paramagnetic relaxation with dissolved oxygen;
    (*iii*) spin–rotation relaxation; or
    (*iv*) anisotropic chemical shift or other mechanisms as appropriate.

We shall not usually be concerned with the precise composition of ρ^* but, in general, it cannot be neglected.

B. The Cross-Correlation Problem

The derivation of Eq. (3.1) makes an important and not entirely justified assumption—the effects of *cross-correlation* have been neglected. Dipole–dipole relaxation depends on the random motions of the vector connecting the two nuclear dipoles involved with respect to the external applied magnetic field. In a rigid molecule, the vectors between different pairs of nuclei do not diffuse independently; to the contrary, their motions must be correlated with each other. For example, in a three-spin system the conditional probability

$$P(\mathbf{\Omega}_{12}:\mathbf{\Omega}^{\circ}_{12}, \tau)$$

that the vector $\mathbf{r}_{12}$ between spins 1 and 2 has an orientation in space $\mathbf{\Omega}_{12}$ at time τ having had an orientation $\mathbf{\Omega}^{\circ}_{12}$ at $t = 0$ is not independent of

$$P(\mathbf{\Omega}_{13}:\mathbf{\Omega}^{\circ}_{13}, \tau)$$

for the vector $\mathbf{r}_{13}$ because of the fixed relationship between $\mathbf{\Omega}_{12}$ and $\mathbf{\Omega}_{13}$ required by the rigidity of the molecule. Nevertheless, numerous calculations (*1–11*) have shown that the neglect of cross-correlation in various special cases is a reasonable assumption. Most of these calculations are aimed specifically at the effect of cross-correlation on relaxation time measurements and, as such, may not be directly applicable to NOE measurements. The only examination of cross-correlation with respect to NOE experiments is that of Kuhlmann *et al.* (*12*). This study, limited to a single spin relaxed by a group of equivalent spins, concludes that cross-correlation effects are negligible in the weak coupling limit. Other cases were not considered; however, the key step in this derivation, the factoring of the spin density matrix, would seem to be more general than the specific case considered. This may indicate that the derivation could be generalized to sets of weakly coupled nonequivalent spins. However, a general derivation must consider the effects of rapid internal rotation found to be significant by Hubbard (*11*).

The question of the importance of cross-correlation effects in the NOE of multispin systems must still be regarded as open. On the other hand, there is no experimental evidence that they *are* significant, and their neglect is an almost universal practice. We shall likewise assume that the effect of cross-correlation can be neglected.

C. The NOE in Multispin Systems

We use the following conventions: d labels the spin which is observed (*detected*); s labels the spins which are *saturated*; n labels the spins *not* saturated *not* including d. The fractional enhancement of the total integrated intensity of the resonance of spin d when spins s are saturated is defined as

$$f_d(s) = [\langle \mathbf{I}_{zd} \rangle - I_{0d}]/I_{0d} \qquad (3.5)$$

This quantity was called f_s^d in a previous paper (*13*). Note that $f = 0$ corresponds to no change in the signal and $f = -1$ corresponds to total saturation of resonance d. Equation (3.1) can be solved for $f_d(s)$ by setting $\langle \mathbf{I}_{zs} \rangle = 0$ for all of the saturated spins and using the steady-state assumption for spin d

$$d\langle \mathbf{I}_{zd} \rangle/dt = 0$$

This gives:

$$f_d(s) = \sum_s \sigma_{ds} I_{0s}/R_d I_{0d} - R_d^{-1} \sum_n \sigma_{dn}(I_{0n}/I_{0d})(\langle \mathbf{I}_{zn} \rangle - I_{0n})/I_{0n}$$

But $(\langle \mathbf{I}_{zn} \rangle - I_{0n})/I_{0n}$ is simply $f_n(s)$, the enhancement of spin n when s is saturated. Likewise

$$I_{0i} \propto I_i(I_i + 1)\gamma_i$$

and, for dipole–dipole relaxation,

$$\sigma_{ij} \propto (\rho_{ij}/2) I_i(I_i + 1)/I_j(I_j + 1)$$

This gives

$$f_d(s) = \sum_s \gamma_s \rho_{ds}/2\gamma_d R_d - \sum_n \gamma_n \rho_{dn} f_n(s)/2\gamma_d R_d \qquad (3.6)$$

Equation (3.6) is valid for any nuclear spin, but we shall be primarily concerned with spins-$\frac{1}{2}$ where Eqs. (3.3) and (3.4) apply. In particular we shall assume throughout this chapter that $\rho_{ij} = \rho_{ji}$. The sum over s includes all saturated spins; the sum over n includes all spins except d and s. By implication, d is a single spin; if d is one of a group of equivalent spins the enhancements of each member of the group are given by an equation of the form of (3.6) (these enhancements are usually equal). Equation (3.6) with related definitions is the basic equation for the NOE in rigid molecules. As will be seen in the next chapter, it also applies to molecules with internal degrees of freedom under a wide range of circumstances. Because of the dependence of ρ_{ij} on the i–j internuclear distance, the NOE, alone or combined with transient experiments (Chapter 6), is capable of giving information on molecular geometry. However, the naïve presumption that the closer the spins, the larger the NOE enhancement can be wrong for at least two reasons: (a) The NOE depends on R_d^{-1} which will not be the same for all the spins. (b) The second term of Eq. (3.6),

See p 25

which represents the indirect polarization of spin d by s through the other spins n, will often decrease (but occasionally increase) the observed NOE. A more accurate generalization is that the enhancement of a spin d will depend on the *relative* proximity of d to the spins s and n.

Most molecules whose structures are chemically interesting will have a sizable number of spins-$\frac{1}{2}$ which must be included in Eq. (3.6). Technically, spins with $I > \frac{1}{2}$ are also included, but their effect can usually be neglected either because of their small γ, their efficient quadrupole relaxation or, usually, both. Even when a ρ_{ij} to a quadrupolar spin is not totally negligible, its effect can often be included in ρ^* because the terms in σ_{ij} will drop out; this can be seen from Eq. (3.1) when it is realized that efficient relaxation will require $\langle \mathbf{I}_{zj} \rangle \sim I_{0j}$ at all times. For N spins there are $N(N-1)$ enhancements since in general $f_i(j) \neq f_j(i)$, and, although all of these enhancements may not be needed, the mathematical problem can become very complex if $N > 3$. In addition, in a complex NMR spectrum, all of the necessary enhancements may not be measurable—e.g., if there are overlapping resonances. Some simplifications are available. In certain cases, some of the intramolecular dipole–dipole interactions can be lumped with ρ^*; also selective deuteration can simplify some proton spectra. Given a molecule and knowing what information is desired, it is important to be able to predict:

(a) Which NOE enhancements are needed to obtain the desired information and how accurately the enhancements must be known? Frequently more than one set of experiments will give the same information, but one will be more accurate than the others.

(b) Which spin subsystems can be considered alone—lumping the other nuclei with ρ^*? How large are the indirect effects neglected by this procedure?

(c) Can the unmeasurable enhancements be calculated?

(d) Which substitutions of deuterium for protons will prove most useful?

Only experience can answer all of these questions, but it is hoped that the material to follow will be helpful.

D. Two Spins and Two Groups of Equivalent Spins

With only two spins, or two groups of equivalent spins, the steady-state NOE alone cannot directly give structural information. However, such measurements combined with transient NOE experiments (Chapter 6) can sometimes give such information. The formulas of this section can be useful in multispin cases where structural information is directly available from steady-state NOE experiments.

1. Two Spins

For two spins the NOE is simply

$$f_d(s) = \rho_{ds}(\gamma_s/2\gamma_d)/(\rho_{ds} + \rho_d{}^*) \tag{3.7}$$

For homonuclear spins, $f_d(s)$ has a maximum of 50% when $\rho_d{}^*$ is negligible compared with ρ_{ds}.

2. Equivalent spins: ax_2

For equivalent sets we shall call one group a and the other group x; as usual we assume loose coupling so that $J_{ax} \ll \delta_{ax}$.

First, for the enhancement of a when x is saturated, Eq. (3.6) gives directly (since *both* spins x are saturated):

$$f_a(x) = 2(\gamma_x/2\gamma_a)\rho_{ax}/R_a$$

Since ρ_{ax} is the dipole–dipole interaction of a with a single spin x (presumed to be the same for both spins x), it enters twice in R_a, so

$$f_a(x) = 2\rho_{ax}(\gamma_x/2\gamma_a)/(2\rho_{ax} + \rho_a{}^*) \tag{3.8}$$

The maximum possible enhancement is again 50% for homonuclear spins.

To calculate $f_x(a)$ from Eq. (3.6) we must consider one of the spins x to be spin d and the other to be spin n. However, if ρ_{ax} (or at least its average) is the same for both spins, their enhancements will be the same. This gives

$$f_x(a) = (\gamma_a/2\gamma_x)\rho_{ax}/R_x - \gamma_x\rho_{xx}f_x(a)/2\gamma_x R_x$$

with

$$R_x = \rho_{xx} + \rho_{xa} + \rho_x{}^* \quad \text{and} \quad \rho_{xx} = \gamma_x{}^4\hbar^2\tau_c(xx)/r_{xx}^6 \tag{3.9}$$

as a special case of Eq. (3.4). Note that there is no $\frac{1}{2}$ factor for identical spins in the definition of ρ_{xx}. Solving for f, we get

$$f_x(a) = (\gamma_a/2\gamma_x)\rho_{ax}/(\tfrac{3}{2}\rho_{xx} + \rho_{ax} + \rho_x{}^*) \tag{3.10}$$

It can be seen that the "$\frac{3}{2}$ effect" is automatically accounted for by this method.

3. Other Special Cases

By an exactly analogous method the following formulas can be derived

ax_3:

$$f_a(x) = 3\rho_{ax}(\gamma_x/2\gamma_a)/(3\rho_{ax} + \rho_a^*) \tag{3.11}$$

$$f_x(a) = \rho_{xa}(\gamma_a/2\gamma_x)/(3\rho_{xx} + \rho_{xa} + \rho_x^*) \tag{3.12}$$

a_2x_2:

$$f_a(x) = 2\rho_{ax}(\gamma_x/2\gamma_a)/(\tfrac{3}{2}\rho_{aa} + 2\rho_{ax} + \rho_a^*) \tag{3.13}$$

$f_x(a)$ can be obtained from Eq. (3.13) by exchanging x and a.

a_2x_3:

$$f_a(x) = 3\rho_{ax}(\gamma_x/2\gamma_a)/(\tfrac{3}{2}\rho_{aa} + 3\rho_{ax} + \rho_a^*) \tag{3.14}$$

$$f_x(a) = 2\rho_{xa}(\gamma_a/2\gamma_x)/(3\rho_{xx} + 2\rho_{xa} + \rho_x^*) \tag{3.15}$$

4. The General Case

Consider two groups of equivalent spins, N_i spins i and N_j spins j. If we assume that, on the average, ρ_{ij} is the same for all i–j pairs, Eq. (3.6) gives

$$f_i(j) = N_j\rho_{ij}(\gamma_j/2\gamma_i)/R_i - \tfrac{1}{2}(N_i - 1)\rho_{ii}f_i(j)/R_i$$

We define

$$R_{ii} = R_i + \tfrac{1}{2}(N_i - 1)\rho_{ii} \tag{3.16}$$

This will later be identified as the total direct relaxation rate of a group of equivalent spins i. If only dipole–dipole interactions with the spins j (and the other spins i) need be considered,

$$R_{ii} = \tfrac{3}{2}(N_i - 1)\rho_{ii} + N_j\rho_{ij} + \rho_i^* \tag{3.17}$$

and

$$f_i(j) = N_j\rho_{ij}\gamma_j/2\gamma_i R_{ii} \tag{3.18}$$

The close similarity between Eqs. (3.18) and (3.7) should be noted. However, in the homonuclear case, the maximum NOE enhancement of 50% seems less likely to be possible for groups of equivalent spins since it requires that both ρ_{ii} and ρ_i^* be negligible compared to ρ_{ij}.

5. Relaxation Times: The $\tfrac{3}{2}$ Effect

As stated previously the generalization that R_d corresponds to $1/T_{1d}$ does not apply to equivalent spins because it does not take the "$\tfrac{3}{2}$ effect" into proper account. If we have N_i spins i only the total I_z of these spins

$$\langle \mathbf{I}_z^{\mathrm{T}} \rangle = \sum_i \langle \mathbf{I}_{zi} \rangle \tag{3.19}$$

can be observed. In Eq. (3.1) for one of the spins i, the second term contains $N_i - 1$ terms in σ_{ii}. When all of the Eqs. (3.1) for the equivalent group are summed, one obtains

$$d\langle \mathbf{I}_z^\mathsf{T} \rangle / dt = -[R_i + (N_i - 1)\sigma_{ii}] [\langle \mathbf{I}_z^\mathsf{T} \rangle - I_0^\mathsf{T}] - N_i \sum_j{}' \sigma_{ij} [\langle \mathbf{I}_{zj} \rangle - I_{0j}] \tag{3.20}$$

The prime on the sum indicates that all spins i have been excluded from the sum. Thus the total direct relaxation rate of the spins i is

$$R_{ii} = R_i + (N_i - 1)\sigma_{ii} \tag{3.21}$$

This definition is slightly more general than Eq. (3.17); however, if only dipole–dipole relaxation contributes to σ_{ii} (no chemical exchange), Eqs. (3.21) and (3.17) are identical.

Thus, for a group of equivalent spins i, the spin–lattice relaxation time T_{1i} corresponds, roughly, to R_{ii}^{-1} and the "$\frac{3}{2}$ effect" is properly accounted for. The correspondence is exact when $R_{ii} \gg$ all σ_{ij}.

6. Nearly Equivalent Spins

Consider the four spin case $abcx$ where a, b, and c are tightly coupled and x is loosely coupled to spins a, b, and c. The simplest case is if all three tightly coupled spins are saturated:

$$f_x(a, b, c) = (\rho_{ax} + \rho_{bx} + \rho_{cx})(\gamma_a / 2\gamma_x) / R_x \tag{3.22}$$

which is the same as for equivalent spins. If in addition we have

$$\rho_{ax} \sim \rho_{bx} \sim \rho_{cx} \quad \text{and} \quad \rho_{ab} \sim \rho_{ac} \sim \rho_{bc}$$

the enhancement of the total NMR resonance of all the tightly coupled spins when x is saturated, $f_{abc}(x)$, will be given by the preceding equations for equivalent spins. These simplifications are frequently useful when tightly coupled spins cannot be avoided. Other effects of tight coupling will be discussed in Appendix I.

7. Internuclear Distances from Two-Spin NOE

Bell and Saunders (14) have described a method for determining the distance between two protons directly from the NOE enhancement. Their method rests on the assumption that ρ^* is due entirely to intermolecular dipole–dipole relaxation with the solvent and is the same for all molecules in a given solvent. Rearranging Eq. (3.7) (with $\gamma_d = \gamma_s$) yields

$$f_d(s)^{-1} = 2 + 2\rho_d^*/\rho_{ds} \tag{3.23}$$

for two protons, and, if s is a methyl group, from Eq. (3.11)

$$f_d(s)^{-1} = 2 + 2\rho_d^*/3\rho_{ds} \tag{3.24}$$

Using Eq. (3.4), these equations can be seen to be of the form

$$f_d(s)^{-1} = 2 + Ar_{ds}^6 \tag{3.25}$$

If it is reasonable to assume that ρ^*/τ_c is a constant, then A in Eq. (3.25) is a constant.

Bell and Saunders measured the NOE enhancements of numerous molecules in carefully degassed and purified $CDCl_3$ solutions at concentrations between 5 and 10% (presumably by weight). The isotopic purity of the chloroform-d used and the amount of TMS added for internal lock were not reported. Their results are shown in Figs. 3-1 and 3-2 for H{H} and H{Methyl} NOE respectively.† Log–log plots of the NOE enhancement versus internuclear distance are linear with a slope of -6 in both cases, indicating that the first term of Eq. (3.25) is negligible.

This method for estimating internuclear distances may be useful under the following circumstances:

(a) The molecules compared are reasonably similar, especially with respect to the observed proton.

(b) The same solvent, TMS concentration, and sample preparation procedure are used in all cases.

(c) The nuclei involved must be neither too close or too distant; ρ^* cannot be too small or too large compared to ρ_{ds}, otherwise the first term of Eq. (3.25) may dominate or the enhancements may be too small to measure.

(d) Intramolecular dipole–dipole interactions must be limited to only the two spins in question.‡

(e) Highly specific solvent–solute interactions should not be present.

---

† The notation A {B} is more or less standard and indicates that spins A are observed while spins B are saturated. It will be used henceforth.

‡ Later we shall see that a zero NOE, $f_i(j) = 0$, does not necessarily imply that $\rho_{ij} = 0$ if more than two spins are interacting (cf. Section H and Fig. 4-1).

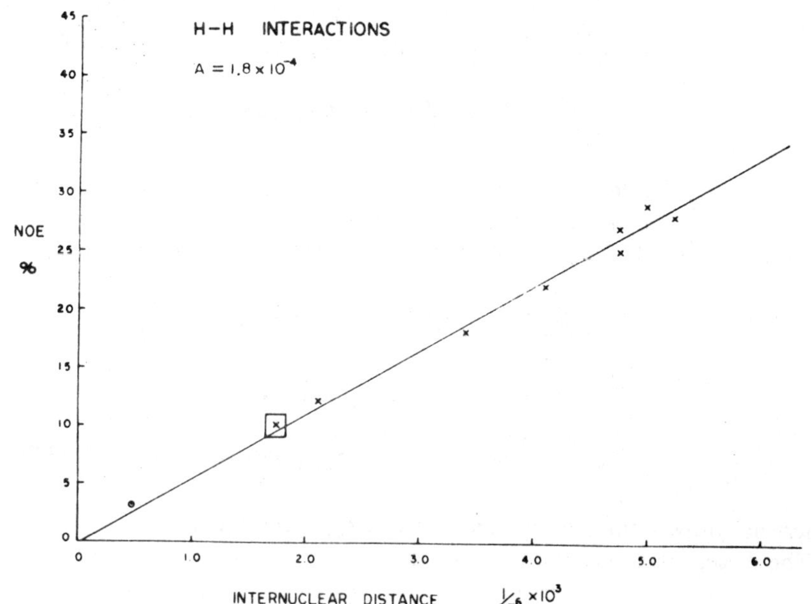

Fig. 3-1   NOE enhancements versus internuclear distance for two-spin H–H interactions. Reproduced by permission of the National Research Council of Canada from R. A. Bell and J. K. Saunders, *Can. J. Chem.* **48**, 1114 (1970).

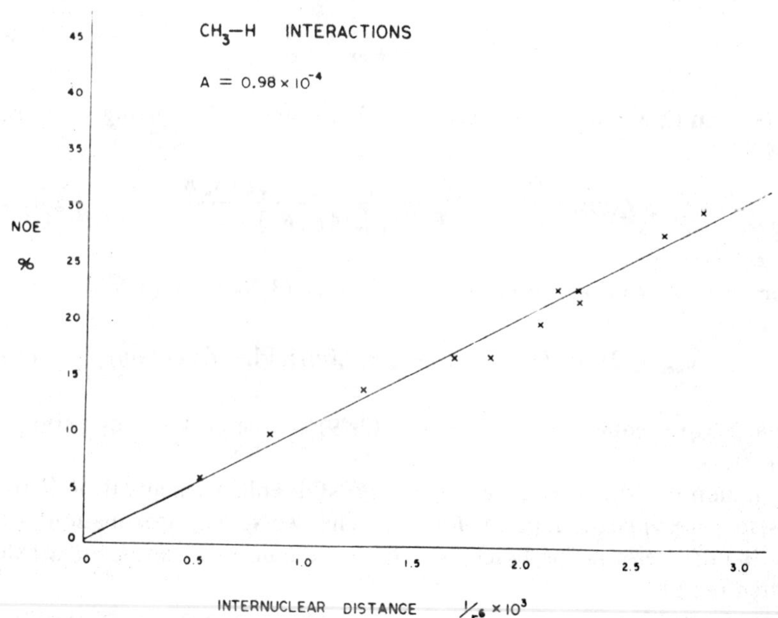

Fig. 3-2   NOE enhancements versus internuclear distance for Me–H interactions. Reproduced by permission of the National Research Council of Canada from R. A. Bell and J. K. Saunders, *Can. J. Chem.* **48**, 1114 (1970).

E. Three Nonequivalent Spins—*amx*

Three nonequivalent spins provide the simplest case for which information on molecular geometry is directly available from steady-state NOE. For three spins, *amx*, there are six NOE enhancements; Eq. (3.6) for three of them is

$$2\gamma_a R_a f_a(m) = \gamma_m \rho_{am} - \gamma_x \rho_{ax} f_x(m) \tag{3.26a}$$

$$2\gamma_a R_a f_a(x) = \gamma_x \rho_{ax} - \gamma_m \rho_{am} f_m(x) \tag{3.26b}$$

$$2\gamma_x R_x f_x(m) = \gamma_m \rho_{mx} - \gamma_a \rho_{xa} f_a(m) \tag{3.26c}$$

There are three similar equations for $f_x(a)$, $f_m(a)$ and $f_m(x)$.

The direct relaxation coefficients are

$$R_a = \rho_{am} + \rho_{ax} + \rho_a^* \tag{3.27a}$$

$$R_m = \rho_{am} + \rho_{xm} + \rho_m^* \tag{3.27b}$$

$$R_x = \rho_{ax} + \rho_{mx} + \rho_x^* \tag{3.27c}$$

$f_x(m)$ can be eliminated between Eqs. (3.26a) and (3.26c) giving

$$f_a(m) = \frac{(\gamma_m/\gamma_a)\,[\rho_{am}/2R_a - \rho_{ax}\rho_{xm}/4R_a R_x]}{[1 - \rho_{ax}^2/4R_a R_x]} \tag{3.28}$$

Similarly ρ_{ax} can be eliminated between Eqs. (3.26a) and (3.26b) giving

$$\rho_{am} = 2R_a(\gamma_a/\gamma_m)\,[f_a(m) + f_a(x)f_x(m)]/[1 - f_m(x)f_x(m)] \tag{3.29}$$

The analogous equations to (3.28) and (3.29) can be obtained by permuting the indices.

Equation (3.28) is useful for estimating NOE enhancements from known, guessed, or estimated values of R and ρ. The two-spin approximation, using only the first term in the brackets in both the numerator and denominator, is often useful.

Equation (3.29) is useful for calculating the fraction of the relaxation of spin a due to spin m, ρ_{am}/R_a. If R_a is measured by another method (Chapter 6), it can be used to measure ρ values if all needed NOE enhancements can be measured.

If R_a cannot be measured conveniently, Eq. (3.29) can be combined with its analog

$$\rho_{ax} = 2R_a(\gamma_a/\gamma_x)\,[f_a(x) + f_a(m)f_m(x)]\,/\,[1 - f_x(m)f_m(x)] \qquad (3.30)$$

to give

$$\frac{\rho_{ax}}{\rho_{am}} = \frac{\gamma_m}{\gamma_x}\frac{[f_a(x) + f_a(m)f_m(x)]}{[f_a(m) + f_a(x)f_x(m)]} \qquad (3.31)$$

If it is safe to assume that $\tau_c(ax) = \tau_c(am)$ (cf. Chapter 2), Eq. (3.31) can be combined with Eq. 3.4 to give

$$\left(\frac{r_{ax}}{r_{am}}\right)^6 = \frac{\gamma_x^{\,3}}{\gamma_m^{\,3}}\frac{f_a(m) + f_a(x)f_x(m)}{f_a(x) + f_a(m)f_m(x)} \qquad (3.32)$$

This case, three loosely coupled nonequivalent spins, is the most complicated for which a closed form equation like (3.32) is convenient. Equation (3.32) is therefore very important, not only for three isolated spins, but as an approximation in spin systems of four or more spins. Unfortunately, most published NOE data is not sufficiently complete to test Eq. (3.32). Of particular interest is its accuracy in determining the ratio of the internuclear distances. Before examining the limited data available, it is worth noting that the sixth power dependence of the left-hand side will be quite helpful—the error in r_{ax}/r_{am} will be approximately one-sixth that of the right-hand side of Eq. (3.32). The assumption that the correlation times are the same must be examined for each specific case; it will probably be reasonable in most large and unsymmetrical molecules.

Already it is obvious that, if the location of a is sought, the most important NOE enhancements are those when a is observed, not those when a is saturated.

It is easy to get confused when transcribing Eq. (3.32) into a real spin system. As a mnemonic, it is worth noting that, for a given ratio r_{ij}/r_{ik}, one spin (i) *must* be in common and the enhancements of that spin appear four times on the right-hand side of Eq. (3.32). Also, the reader will note that a sort of "chain rule" applies for the indices of many of the equations of this chapter; for example, in Eq. (3.32) the first and last indices (a and m) of both terms in the numerator are the same with the intermediate indices occurring in pairs.

F. The *amx₃* Case

The analogs of the above equations for more than three spins are complicated and a more general approach (Section J) is usually indicated. One such case, the five-spin amx_3, is quite similar to the three-spin case and is of special interest because of the common occurrence of methyl groups.

Again, there are six forms of Eq. (3.6) ($\gamma_a = \gamma_m = \gamma_x$):

$$R_a f_a(m) = \tfrac{1}{2}\rho_{am} - \tfrac{3}{2}\rho_{ax} f_x(m) \tag{3.33a}$$

$$R_a f_a(x) = \tfrac{3}{2}\rho_{ax} - \tfrac{1}{2}\rho_{am} f_m(x) \tag{3.33b}$$

$$R_{xx} f_x(a) = \tfrac{1}{2}\rho_{xa} - \tfrac{1}{2}\rho_{xm} f_m(a) \tag{3.33c}$$

$$R_{xx} f_x(m) = \tfrac{1}{2}\rho_{xm} - \tfrac{1}{2}\rho_{xa} f_a(m) \tag{3.33d}$$

$$R_m f_m(a) = \tfrac{1}{2}\rho_{ma} - \tfrac{3}{2}\rho_{mx} f_x(a) \tag{3.33e}$$

$$R_m f_m(x) = \tfrac{3}{2}\rho_{mx} - \tfrac{1}{2}\rho_{ma} f_a(x) \tag{3.33f}$$

with $R_{xx} = R_x + \rho_{xx}$.

From Eqs. (3.33a) and (3.33d) we obtain

$$f_x(m) = \frac{\rho_{xm}/2R_{xx} - \rho_{xa}\rho_{am}/4R_a R_{xx}}{1 - 3\rho_{xa}\rho_{ax}/4R_a R_{xx}} \tag{3.34a}$$

From Eqs. (3.33b) and (3.33f) we obtain

$$f_m(x) = \frac{3\rho_{mx}/2R_m - 3\rho_{ma}\rho_{ax}/4R_a R_m}{1 - \rho_{ma}\rho_{am}/4R_a R_m} \tag{3.34b}$$

From Eqs. (3.33a) and (3.33d) we obtain

$$f_a(m) = \frac{\rho_{am}/2R_a - 3\rho_{ax}\rho_{xm}/4R_a R_{xx}}{1 - 3\rho_{ax}\rho_{xa}/4R_a R_{xx}} \tag{3.34c}$$

Three similar equations can be obtained by exchanging the indices a and m. Equation (3.34c) is interesting. R_{xx} is frequently relatively large for methyl groups, so the effect of indirect polarization on $f_a(m)$ or $f_m(a)$ due to methyl groups (the negative terms in the numerator and denominator) will be relatively small, especially if a and m are closer to each other than to x_3. Thus, methyl groups are prime candidates for inclusion in ρ^*. A similar argument can be applied to quadrupolar and other nuclei which relax rapidly. Similarly, enhancements when methyl groups are observed [Eq. (3.34a)] will usually be small.

It is more useful to solve Eqs. (3.33) to find ρ in terms of f. From Eqs. (3.33a) and (3.33b)

$$\rho_{am} = \frac{2R_a[f_a(m) + f_a(x)\,f_x(m)]}{[1 - f_m(x)\,f_x(m)]} \tag{3.35a}$$

From Eqs. (3.33c) and (3.33d)

$$\rho_{xa} = \frac{2R_{xx}[f_x(a) + f_x(m)\,f_m(a)]}{[1 - f_a(m)\,f_m(a)]} \tag{3.35b}$$

Curiously, another equation like (3.35b) is available from Eqs. (3.33a) and (3.33b)

$$\rho_{ax} = \frac{2R_a}{3}\frac{[f_a(x) + f_a(m)\,f_m(x)]}{[1 - f_x(m)\,f_m(x)]} \tag{3.35c}$$

A similar pair of equations could have been obtained from the *amx* case, but then they would have been trivially identical—differing only by an exchange of the *a–x* indices. In the equations of this section, *a* and *m*, not *x*, can be exchanged to obtain new equations.

Obviously, equations of the form of (3.31) are available; one of these is

$$\frac{\rho_{am}}{\rho_{ax}} = 3\,\frac{f_a(m) + f_a(x)\,f_x(m)}{f_a(x) + f_a(m)\,f_m(x)} \tag{3.36}$$

The factor of three, which did not occur in Eq. (3.31), occurs here because we have defined ρ_{ax} to be the relaxation coupling of spin *a* with one of the spins *x*. If we solved the preceding equations for ρ_{ax}/ρ_{mx}, an equation exactly analogous to Eq. (3.31) would be obtained because, with the spins *x* in common, the factor of three cancels.

For heteronuclear spins, the γ factors can be inserted into the preceding equations by replacing all $f_i(j)$ with $\gamma_i f_i(j)/\gamma_j$.

It is worth reemphasizing that ρ_{ax} is the dipole–dipole coupling of spin *a* with *one* spin *x* and has been assumed to be equal for all spins *x*. If the ρ_{ax} are made equal by the averaging effect of rapid internal rotation, the distance dependence of Eq. (3.4) becomes $\langle r_{ax}^{-6}\rangle$; the limitations of this statement will be explored in detail in Chapter 4.

G. Three Spins: Linear

We shall now consider three spins with geometry such that $r_{13} > r_{12} \sim r_{23}$ and the 1–2–3 angle is obtuse. The mathematical criterion for such a case is that $\rho_{13} \ll \rho_{12}$ or ρ_{23} rather than any strict limitation on the angle. We shall assume throughout that $\gamma_1 = \gamma_2 = \gamma_3$.

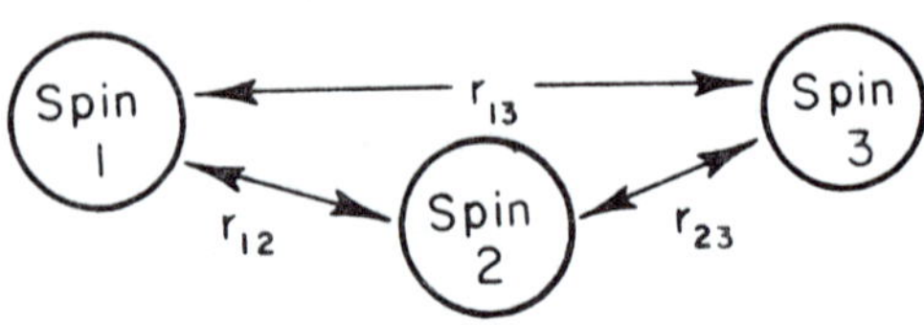

We shall now consider the several types of experiments possible in such a system and approximate expressions for their results.

1. The Central Spin Is Saturated

For the experiment 1{2},

$$f_1(2) = \rho_{12}/2R_1 - \rho_{13}f_3(2)/2R_1$$

If we neglect the term in ρ_{13}, the enhancements can be approximated by the two spin formulas

$$f_1(2) \sim \rho_{12}/2R_1 \tag{3.37a}$$

$$f_3(2) \sim \rho_{32}/2R_3 \tag{3.37b}$$

2. The Central Spin Is Observed

Taking into account the indirect effect through spin 3 for 2{1} but neglecting terms in ρ_{13}

$$f_2(1) = \rho_{21}/2R_2 - (\rho_{23}/2R_2)f_3(1)$$

and

$$f_3(1) \simeq - (\rho_{32}/2R_3)f_2(1)$$

therefore,

$$f_2(1) \simeq (\rho_{21}/2R_2)/(1 - \rho_{23}^2/4R_2 R_3) \tag{3.38a}$$

The negative term in the denominator is often small compared to one so the two-spin approximation [Eq. (3.37)] is sometimes useful. If we use this approximation for $f_2(3)$ and $f_3(2)$, Eq. (3.38a) becomes

$$f_2(1) \simeq (\rho_{21}/2R_2)/[1 - f_3(2)f_2(3)] \tag{3.38b}$$

and

$$f_2(3) \simeq (\rho_{23}/2R_2)/[1 - f_1(2)f_2(1)] \tag{3.38c}$$

Equations (3.38b) and (3.38c) use the two-spin approximation for spin 2 and therefore are less accurate than (3.38a); however, they can be used iteratively and do furnish an experimental test for the validity of the two-spin approximation for a central spin. It may appear from Eqs. (3.38) and (3.37) that $f_2(1)$ will be larger than $f_1(2)$ because of the denominator; however, since R_2 will usually be larger than R_1 (because spin 2 is nearer to spin 3) the opposite will more often be true.

3. The Three-Spin Effect

If an end spin is observed while the other end spin is saturated, using

$$f_3(2) \simeq \rho_{23}/2R_3$$

we obtain

$$f_3(1) \simeq -f_3(2)f_2(1) \tag{3.39a}$$

$$f_1(3) \simeq -f_1(2)f_2(3) \tag{3.39b}$$

Thus in the linear case enhancements two spins away will be *negative*. Such a case was observed by Bell and Saunders (*15*) in *ochotensimine*. Equations (3.39) are useful for estimating such effects and give an upper limit to how negative $f_1(3)$ and $f_3(1)$ can be since a positive term, $\rho_{13}/2R_3$ and $\rho_{13}/2R_1$, respectively, has been neglected. Such negative enhancements are characteristic of "linear" spin systems; and, in fact, form a useful experimental criterion for classifying spin systems as "linear." Experimental definition: a "linear" spin configuration is one in which sizable negative enhancements occur when the second spin away from the saturated spin is observed.

4. Multiple Saturation

If spin 2 is observed when both 1 and 3 are saturated

$$f_2(1 + 3) = (\rho_{12} + \rho_{23})/2R_2$$

To the extent that the two spin approximation is valid when spin 2 is observed [$f_1(2)f_2(1)$ and $f_3(2)f_2(3)$ are small compared with one]

$$f_2(1 + 3) \simeq f_2(1) + f_2(3) \tag{3.40}$$

Thus, the effect of multiple resonance NOE is approximately the sum of the enhancements when one spin at a time is saturated. This generalization works best if the observed spin is centrally located with respect to the saturated spins.

5. A Contrived Example

If we take

$$r_{12}:r_{23}:r_{13} = 1.00:1.10:2.00$$

giving

$$\rho_{12}:\rho_{23}:\rho_{13} = 1.00:0.565:0.015$$

and assume all $\rho^* = 0$ (not likely in a realistic case) we get (in units making $\rho_{12} = 1.000$)

$$R_1 = 1.01, \quad R_2 = 1.565, \quad R_3 = 0.58$$

First we calculate the approximate enhancements using Eq. (3.37) when 2 is saturated, Eq. (3.38) when 2 is observed, and Eq. (3.39) for the three spin enhancements. The results are tabulated on Table 3-1.

TABLE 3-1

APPROXIMATE NOE ENHANCEMENTS FOR $r_{12}:r_{23}:r_{13} = 1.00:1.10:2.00$
ASSUMING $\rho^* = 0$ FOR ALL SPINS

Observe	{1}	{2}	{3}
1	—	0.49	−0.12
2	0.35	—	0.21
3	−0.17	0.49	—

Accurate calculations using Eq. (3.38) give

$$f_3(1) = -0.157 \quad \text{(compare } -0.17\text{)}$$

$$f_3(2) = +0.481 \quad \text{(compare } 0.49\text{)}$$

The others are within 0.01 of the values given on the table. These enhancements represent a maximum effect since the ρ^* were assumed to be zero. Several points:

(a) The difference between the estimated and actual $f_1(3)$ (about 0.01) is due to ρ_{13} being nonzero. With this difference being so small extremely accurate measurements would be needed to determine r_{13}; this will be discussed further below.

(b) The equality of the enhancements of spins 1 and 3 when spin 2 is saturated does not mean that spins 1 and 3 are equidistant from 2; the inequality of the enhancements of spin 2 when 1 and 3 are saturated is more indicative of the distance ratio. Indeed, assuming that ρ_{21}/ρ_{23} is equal to $f_2(1)/f_2(3)$

would, in this case, introduce an error of only a few percent in the ratio r_{12}/r_{23}.

(c) The approximate formulas of this section may be useful in estimating missing or unmeasurable enhancements. One should bear in mind, however, that the errors introduced are systematic, not random, and estimated enhancements should be used cautiously for structural determinations.

6. A Real Example: Ochotensimine

Bell and Saunders (15) have reported considerable data for the NOE of *ochotensimine* (Fig. 3-3). Four of the protons (15A, 15B, 13, and 12) are sufficiently isolated from the rest of the molecule to be treated alone. Furthermore, these four spins are in a "linear" configuration so that the three spin equations of this section can be used as an approximation (cf. Section I).

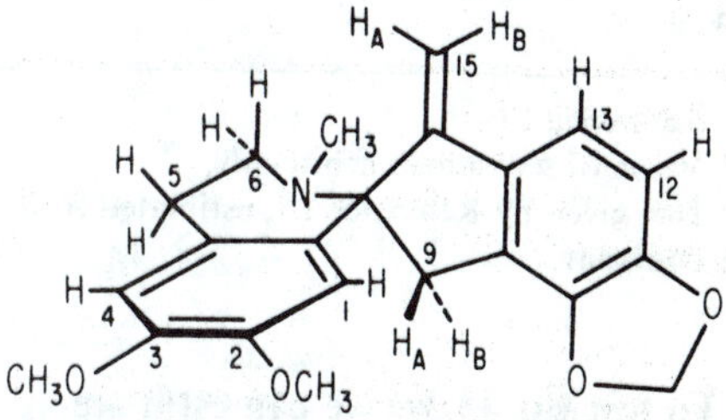

Fig. 3-3 The structure of *ochotensimine*. Reproduced with permission from Bell and Saunders (15).

Unfortunately, several critical enhancements either were not or could not be measured; therefore, we will use the equations of this section to estimate the missing enhancements. The results of Bell and Saunders together with the calculated missing enhancements are given in Table 3-2.

TABLE 3-2

NOE Enhancements for Four Protons of *Ochotensimine* Measured on a Carefully Degassed 0.17 M Solution in $CDCl_3$ with TMS Added as an Internal Reference[a]

Observe	{15A}	{15B}	{13}	{12}
15A	—	0.40	$(-0.03)^b$	$(0.01)^d$
15B	0.40	—	0.08	$(-0.01)^c$
13	-0.08	0.24	—	$(0.17)^c$
12	0.03	-0.07	$(0.29)^b$	—

[a] Reference *15*.

[b] Estimated using Eq. (3.39).

[c] Estimated assuming $\rho_{12,13} = 0.70\,\rho_{13,15B}$.

[d] Estimated in Section I below, Eq. (3.43).

In addition, Bell and Saunders measured the approximate internuclear distances with a Dreiding molecular model. These distances, together with relative values of ρ estimated from them (with respect to $\rho_{15B,13} = 1.00$ assumed) are listed in Table 3-3.

TABLE 3-3

VALUES OF INTERNUCLEAR DISTANCES IN *Ochotensimine* AS GIVEN BY BELL AND SAUNDERS.[a] RELATIVE VALUES OF THE CORRESPONDING ρ'S ARE GIVEN IN PARENTHESIS.

	15B	13	12
15A	1.8 Å (5.64)	4.0 Å (0.05)	6.2 Å (0.00)
15B	—	2.4 Å (1.00)[b]	4.8 Å (0.02)
13	—	—	2.55 Å (0.70)[c]

[a] Reference *15*.

[b] Value of ρ assumed arbitrarily.

[c] Not given by Reference *15*; estimated from the other distances assuming all four spins are coplanar.

To test Eq. (3.39) we can estimate

$$f_{13}(15A) \simeq -f_{13}(15B)f_{15B}(15A) = -9.6\%$$

Reference *15* does not give uncertainties for their measured enhancements but the -9.6% value is probably within experimental error of the reported enhancement, -8%.

Since this value was reasonable, we use Eq. (3.39) to estimate

$$f_{15A}(13) \simeq -f_{15A}(15B)f_{15B}(13) = -0.03$$

$$f_{12}(13) \simeq -f_{12}(15B)/f_{13}(15B) = 0.29$$

We now have sufficient data to use Eqs. (3.31) and (3.32) to calculate relative internuclear distances

$$\frac{\rho_{15B-15A}}{\rho_{15B-13}} = \frac{f_{15B}(15A) + f_{15B}(13)f_{13}(15A)}{f_{15B}(13) + f_{15B}(15A)f_{15A}(13)}$$

$$= \frac{0.40 - 0.0064}{0.08 - 0.012} = 5.78$$

It can be seen that the error in estimating $f_{15A}(13)$ will have a negligible effect on the answer. Taking $r_{15A-15B} = 1.8\text{Å}$ as given in Table 3-3 we obtain

$$r_{15B-13} = (1.8)(5.78)^{1/6} = 2.4\text{Å}$$

This is exactly the distance measured on the model (Table 3-3).

It is impossible to calculate the 12–13 distance, since the enhancements for 13{12} and 15B{12} were not reported. On the other hand, we do have enough data to attempt to calculate the 13–15A distance.

$$\frac{\rho_{13-15B}}{\rho_{13-15A}} = \frac{f_{13}(15B) + f_{13}(15A)f_{15A}(15B)}{f_{13}(15A) + f_{13}(15B)f_{15B}(15A)} = \frac{0.24 - 0.032}{-0.08 + 0.096} = 13.0$$

This value cannot be trusted because the denominator is a difference less than the probable experimental error [Eq. (3.39), of course, predicts the difference to be zero!]. Even so we obtain

$$r_{13,15A} = (2.4)(13.0)^{1/6} = 3.7\text{Å} \qquad (4.0 \text{ reported})$$

which is a surprisingly good result. If $f_{13}(15B)$ has an error as small as 0.01, $\rho_{13-15B}/\rho_{13-15A}$ varies from 8 to 35! Thus, all we can say with certainty is that the results are consistent with a 13–15A distance of 3.4 to 4.35 Å. On the other hand, an error of 0.01 in $f_{13}(15B)$ would result in a much smaller error— the 13–15A distance varies from 3.6 to 3.9 Å in this case.

The use of Eqs. (3.31) and (3.32) for nonadjacent spins is risky and the errors are large.

TABLE 3-4

RELAXATION RATES OF PROTONS IN *Ochotensimine*
RELATIVE TO $\rho_{15B,13} = 1.00$.

Proton	15A	15B	13	12
R	6.9	7.0	2.0	1.2
ρ^*	1.2	0.34	0.25	0.5

^{a} ρ^* values include intramolecular relaxation to other protons in the molecule not included in this group.

Using Eq. (3.29) we can calculate relative values of R_{15A}, R_{15B}, and R_{12}

from the observed enhancements. R_{13} cannot be thus calculated because several enhancements are still missing; however, it will be approximately

$$R_{13} \simeq \rho_{13,15B}/2f_{13}(15B)$$

Given all of the R values, the remaining missing enhancements can be estimated (Table 3-2, values footnoted b). When all of the ρ_{ij} values (Table 3-3) are summed ρ^* can be calculated. These values of R and ρ^* are given in Table 3-4 relative to $\rho_{13,15B} = 1.00$ (i.e., in multiples of $\rho_{13,15B}$).

Considering the experimental error, the ρ^* values for protons $15B$, 13, and 12 are, for all practical purposes, nearly equal; this is as expected for intermolecular relaxation. On the other hand, ρ^* for $15A$ is relatively large; this is presumably because of intramolecular relaxation between H–15A, the nearby N–methyl protons and the protons on C–6 which were not included in the analysis. Indeed, a small (2%) enhancement was observed for 15A{NMe}; no enhancement was reported for $15A$\{H–6\}.

It may seem surprising at first that the mutual enhancements of the 15A and 15B protons are equal. It can now be seen that the $15A\{15B\}$ enhancement is reduced by the "extra" ρ^* of the 15A proton, and the $15B\{15A\}$ enhancement is reduced by the relaxation effect on 15B of the nearby 13 proton. Thus, the equality is accidental. It can also be seen that a 40% enhancement for the NOE of geminal protons is quite reasonable without invoking scalar relaxation as suggested by Reference 15.

7. Conclusion

The "linear" three-spin system is a favorable geometry because the enhancements are large and can be measured accurately. Relative distances between adjacent spins can be determined easily and accurately. "Linear" four- or more spin systems can be approximated well with the three-spin equations of this section. The most important enhancements for determining the ratio r_{12}/r_{32} where spin 2 is between spins 1 and 3 are those when spin 2 is observed. Enhancements between next-nearest neighbors will be small and often negative in this geometry; distances between next-nearest neighbors cannot be determined accurately.

H. Three Spins: Acute Case

We consider now the case where the 1–2–3 angle is acute, including the case when it is on the order of 90°.

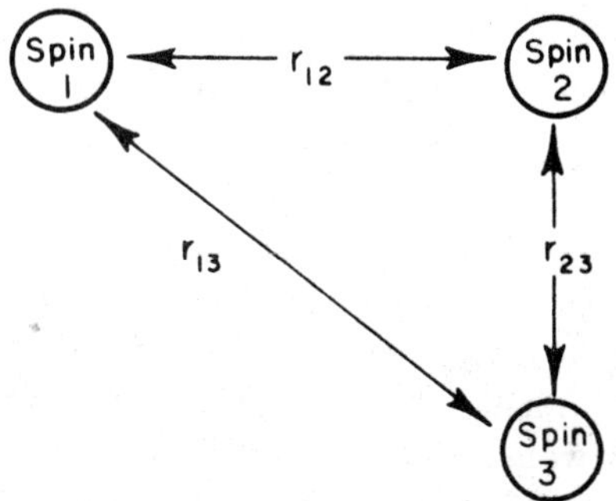

We shall still number the spins so that $r_{13} > r_{12}$, or r_{23}, but consider the case when ρ_{13} is not negligible with respect to ρ_{12} and ρ_{23}. For this reason, the convenient approximations of Section G are no longer valid and the exact equations of Section E must be used.

1. A Contrived Example: Zero NOE for Adjacent Spins

We shall now consider a hypothetical structure with

$$r_{12}:r_{23}:r_{13} = 1.00:1.10:1.30$$

which gives

$$\rho_{12}:\rho_{23}:\rho_{13} = 1.00:0.56:0.21$$

Again we arbitrarily assume $\rho_{12} = 1.000$ so that all other relaxation parameters are in multiples of ρ_{12}. We likewise assume that all $\rho^* = 0$. This gives: $R_1 = 1.21$, $R_2 = 1.56$, and $R_3 = 0.77$. The enhancements calculated with Eq. (3.28) are shown in Table 3-5.

TABLE 3-5

NOE ENHANCEMENTS FOR $r_{12}:r_{23}:r_{13} = 1.00:1.10:1.30$
ASSUMING ALL $\rho^* = 0$

Observe	{1}	{2}	{3}
1	—	0.39	−0.004
2	0.32	—	0.22
3	+0.02	0.31	—

It is particularly relevant to show the details of the calculation of $f_1(3)$:

$$f_1(3) = \rho_{13}/2R_1 - \rho_{12}f_2(3)/2R_1$$
$$= 0.087 - 0.091 = -0.004$$

For $f_3(1)$,

$$f_3(1) = \rho_{31}/2R_3 - \rho_{32}f_2(1)/2R_3$$
$$= 0.136 - 0.116 = +0.02$$

In other words, the 1{3} and 3{1} enhancements *are small not because 1 and 3 are distant but because sizable direct positive and indirect negative effects cancel!* This phenomenon is particularly relevent to the method of Bell and Saunders described in Section D. Smaller 1–2–3 angles would give positive enhancements and larger angles would give negative effects.

Compared to the "linear" case, the enhancements of the acute case are smaller. In the limiting case of three spins on an equilateral triangle, all of the enhancements approach 20% as a maximum. All of these can be greatly reduced by ρ^*. The smaller enhancements combined with typical experimental errors will decrease accuracy in the acute angle case. On the other hand, for smaller angles both ratios r_{12}/r_{23}, r_{13}/r_{23} may be obtained, albeit perhaps with poor accuracy, while in the "linear" case r_{13} often cannot be measured at all.

As an experimental criterion, we define an "acute" configuration as one in which $f_1(3)$ and $f_3(1)$ are small or positive.

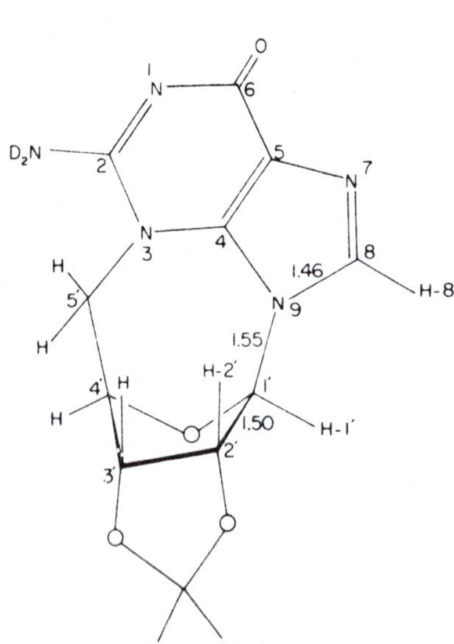

Fig. 3-4 The structure of 2′,3′-isopropylidene-3,5′cycloguanosine (i-cG). [Reprinted by permission from R. E. Schirmer, J. H. Noggle, J. P. Davies, and P. A. Hart, *J. Amer. Chem. Soc.* **92**, 3266 (1970).]

2. *An Actual Example: 2′,3′-isopropylidene-3,5′-cycloguanosine (i-cG)*

The investigation of the NOE for the protons of i-cG by Schirmer, Noggle, Davis, and Hart (*13*) furnishes a good example of an acute angle case (the

TABLE 3-6

EXPERIMENTAL NOE ENHANCEMENTS FOR 2', 3'-ISOPROPYLIDENE-
3, 5'-CYCLOGUANOSINE (i–cG) AS REPORTED BY SCHIRMER *et al.*[a]

Observe	{1'}	{2'}	{8}	{3'}
1'	—	0.10	0.32[b]	0.05
2'	0.10	—	0.00	—
8	0.36	0.00	—	0.00

[a] See Schirmer *et al.* (*13*). The solution was 0.25 M in DMSO-d_6.

[b] The experimental value was actually 0.27 but 19% of H–8 had been exchanged for ^{2}D and the value reported was corrected for this effect.

angle is actually quite near 90°). The structure of i-cG is shown in Fig. 3-4. An NOE enhancement of 5% was observed between H1'{H3'}, because the 3' resonance was near the 2' resonance, 2'{3'} was not measured. However, since 3' is on the outside of the 1–2–8 triangle and relatively distant from 1' and 8, we will include its effect in ρ^* and treat 1'–2'–8 as a three-spin system.

The experimental NOE enhancements are shown in Table 3-6. The internuclear distances measured on a molecular model ("Framework Molecular Models." Prentice-Hall, Inc.) and calculated from Eq. (3.32) are given in Table 3-7.

TABLE 3-7

INTERPROTON DISTANCES IN i–cG[a]

Protons	$\phi = 0°$ [b]		$\phi = 35°$ [b]	
	Measured on model	Equation (3.32)	Measured on model	Equation (3.32)
1'–2'	2.8	c	2.9	c
1'–8	2.5	2.3	2.6	2.4
2'–8	3.7	3.4	4.6	3.5

[a] These numbers differ slightly from those reported by Schirmer *et al.* (*13*) due to an arithmetical error in the original reference.

[b] ϕ is the angle between the glycosidic bond and its projection on the plane of the base, cf. Reference *13*.

[c] The value of the 1'–2' distance is taken as given.

The smaller distance, $1'$–8, in i-cG is determined less accurately than in the previous example. The error can be estimated from

$$(r_{1'2}/r_{1'8})^6 \sim f_{1'}(8)/f_{1'}(2') = 0.32/0.10$$

The primary source of error is the small and difficult to measure $f_{1'}(2')$; the reported uncertainty of ± 0.05 in this value would, if $f_{1'}(8)$ is assumed to be errorless, indicate an error of about 8% (one-sixth of the percentage of error in ρ) for the $1'$–8 distance so

$$f_{1'8} = 2.3 \pm 0.2\,\text{Å}$$

This is a reasonable uncertainty (cf. Table 3-7).

For the longer $2'$–8 distance, the situation is somewhat better than for the linear case

$$\left(\frac{r_{1'8}}{r_{2'8}}\right)^6 = \frac{f_8(2') + f_8(1')f_{1'}(2')}{f_8(1') + f_8(2')f_{2'}(1')} \simeq f_{1'}(2')$$

since $f_8(2') = 0$. If the error is due only to $f_{1'}(2')$, the error in the $2'$–8 distance is another 8% on top of the error in the $1'$–8 distance. (Of course if $r_{1'8}$ were taken as given only 8% error would enter into both the $1'$–$2'$ and $1'$–8 distances.) On the other hand, the equation

$$\left(\frac{r_{2'8}}{r_{1'2'}}\right) = \frac{f_{2'}(1') + f_{2'}(8)f_8(1')}{f_{2'}(8) + f_{2'}(1')f_{1'}(8)} \sim f_{1'}(8)^{-1}$$

[since $f_{2'}(8) = 0$] seems to be more accurate since $f_{1'}(8)$ was larger and could be measured more accurately. However, if the zero $2'\{8\}$ enhancement was as large as 0.015, the error would again be 8% or

$$r_{1'2'} = 3.4 \pm 0.3$$

The uncertainties in the "zero" enhancement are probably larger than this so the above error estimate is certainly conservative.

The difficulty with i-cG is twofold:

(a) The relaxation times of the protons were short, with $T_1 \sim$ a few seconds (*16*); this necessitated large power levels to saturate the spins and made the experiment difficult to do.

(b) The geometry causes the enhancements to be quite small so that a fixed error in the intensity measurement is more important than when the enhancements are larger.

Again, relative values of the relaxation parameters can be estimated (in multiples of $\rho_{1'8}$):

$$\rho_{2'8} = 0.10, \qquad \rho_{1'2'} = 0.31$$

$$\rho_8{}^* = 0.28, \qquad \rho_{1'}^* = 0.25$$

($\rho_{1'}^*$ includes the 1'–3' interaction.)

$$R_8 = 1.38, \qquad R_{1'} = 1.59$$

The values of ρ^* and R for H–2' cannot be estimated from the data given.

3. Conclusion

In the three spin acute angle case it is possible to determine both ratios of internuclear distances; however, the small enhancements typically obtained in this case are difficult to measure with sufficient accuracy. Because of cancelation of terms in Eq. (3.6) it is possible to observe near-zero NOE enhancements between spins whose internuclear distance is *not* so large as to make their dipole–dipole coupling negligible. The convenient approximations such as Eqs. (3.37)–(3.40) are no longer useful.

I. Four Spins

When the number of spins exceeds three, the numerical procedure described in the next section presents a tempting alternative to the use of closed formulas. In addition, in many such cases, the three-spin equations can often be used to good effect. Nevertheless, the four-spin formula is useful as a source of approximations for special cases.

Consider a four-spin system $abcx$ where the location of spin c is sought. (No implication is intended that abc spins are tightly coupled; as usual, all spins must be loosely coupled.) The effect of another spin x can be obtained from

$$\frac{\rho_{ac}}{\rho_{bc}} = \frac{\gamma_b}{\gamma_a}$$

$$\times \frac{f_c(a)+f_c(b)\,f_b(a)+f_c(x)\,f_x(a)+f_c(b)\,f_b(x)\,f_x(a)+f_c(x)\,f_x(b)\,f_b(a)-f_c(a)\,f_b(x)\,f_x(b)}{f_c(b)+f_c(a)\,f_a(b)+f_c(x)\,f_x(b)+f_c(a)\,f_a(x)\,f_x(b)+f_c(x)\,f_x(a)\,f_a(b)-f_c(b)\,f_a(x)\,f_x(a)}$$

$$(3.41)$$

The first two terms in the numerator and denominator of Eq. (3.41) will be recognized as the three-spins formula. For the homonuclear case, all f's are less than one, so the third-order terms of Eq. (3.41) (those in which three f's are multiplied together) will often be small; however, small is a relative word and they must be compared in size to the leading terms which, especially in compact geometries, also will often be small compared to one.

The "linear" four-spin geometry is especially interesting and simple. If the arrangement is $a–c–b–x$ the terms in $f_a(x)$ and $f_x(a)$ can be neglected (see below) giving ($\gamma_a = \gamma_b$) (cf. Table 3-2)

$$\frac{\rho_{ac}}{\rho_{bc}} = \frac{f_c(a) + f_c(b) f_b(a) + f_c(x) f_x(b) f_b(a) - f_c(a) f_b(x) f_x(b)}{f_c(b) + f_c(a) f_a(b) + f_c(x) f_x(b)} \qquad (3.42)$$

For *ochotensimine* (*15*), which we previously approximated as a three-spin case, the fourth term of the numerator and the third term of the denominator are *not* insignificant. However, their effect cancels so Eq. (3.42) gives $\rho_{15A-15B}/\rho_{15-B13} = 5.73$ (cf. 5.78), a totally negligible error.

Now, let us consider a "linear" four spin system with geometry of the type

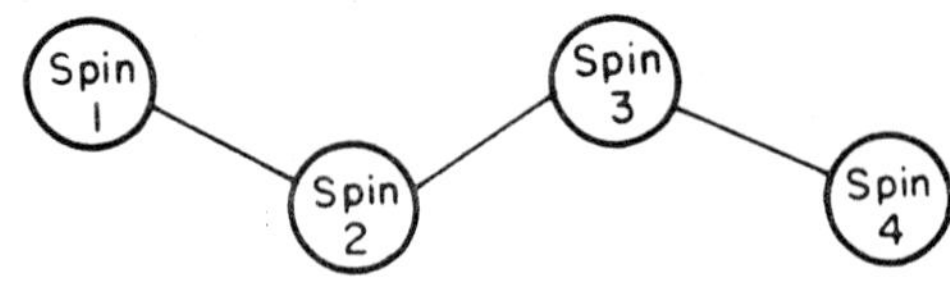

Using Eq. (3.6), we get

$$f_4(1) = \rho_{14}/2R_4 - (\rho_{24}/2R_4) f_2(1) - (\rho_{34}/2R_4) f_3(1)$$

Neglecting the first two terms and using the two-spin approximation

$$\rho_{43}/2R_4 \simeq f_4(3)$$

we get

$$f_4(1) \simeq - f_4(3) f_3(1)$$

Now if we use Eq. (3.39)

$$f_4(1) \simeq f_4(3) f_3(2) f_2(1) \qquad (3.43)$$

Since the most significant correction terms are negative, Eq. (3.43) gives an upper limit for $f_4(1)$. The correction terms are small, but so is the value calculated by Eq. (3.43) so it would not be surprising if $f_4(1)$ were actually zero or slightly negative. In any case, the enhancement of the fourth spin is very small in "linear" cases.

For *ochotensimine* (*15*), we estimate from Eq. (3.43) (cf. Table 3-2)

$$f_{12}(15A) \simeq +0.028$$

remarkably close to the experimental value of $+0.03$. The other four spin enhancement, not reported in Reference *15*, is estimated as

$$f_{15A}(12) \simeq 0.0054$$

—unmeasurably small.

J. Multispin Systems

A convenient form of Eq. (3.6) is obtained if it is first rewritten as

$$2\gamma_d R_d f_d(s) = \sum_s \gamma_s \rho_{ds} - \sum_n \gamma_n \rho_{dn} f_n(s)$$

The two sums on the right-hand side can be combined if we define

$$f_s(s) = -1 \tag{3.44}$$

which is consistent with the definition of f [Eq. (3.5)] since $\langle \mathbf{I}_{zs} \rangle = 0$. Using the definition of R_d [Eq. (3.2)], this equation becomes

$$\sum_{i \neq d} \rho_{di}[f_d(s) + (\gamma_i/2\gamma_d) f_i(s)] + f_d(s)\rho_d{}^* = 0 \tag{3.45}$$

The sum over i includes s but not d. Since an equation of the form of (3.45) exists for each ordered pair of nuclei d–s, there are $N(N-1)$ such equations for N spins. Since there are $\frac{1}{2}N(N+1)$ values of ρ and ρ^* to be determined, the problem is apparently over determined. (Actually, only relative values of ρ and ρ^* can be determined since the equations are homogeneous, so the unknowns number one less than the above.) In practice, not all of the enhancements are measurable and not all of the equations are useful. The problem must frequently be simplified by including some ρ_{ij} terms in ρ^* and supplying some information about the molecule—e.g., that some values of ρ_{ij} are likely to be zero. Generally, if an assumption about the correlation times can be made, Eq. (3.45) is sufficient to determine the relative distances between all nuclei between which an NOE is observed.

For homonuclear spins (all γ's equal) another form of Eq. (3.45) is available if one assumes that all of the correlation times are equal

$$\sum_{i \neq d} \langle r_{di}^{-6} \rangle [f_d(s) + \tfrac{1}{2} f_i(s)] + f_d(s) a_d = 0 \qquad (3.46)$$

$$a_d = \rho_d^* / \gamma_d^4 \hbar^2 \tau_c \qquad (3.47)$$

This form of the equation was used by Schirmer *et al.* (*13*) in their study of 2′,3′-isopropylidene-3,5′-cycloguanosine. Equation (3.46) is particularly convenient for numerical calculations using a computer. If a sufficient set of equations is assembled, the numerical values of the enhancements are entered and, after setting one of the parameters equal to one (most conveniently one of the a_d's), a computer can solve the resulting set of simultaneous linear equations for systems of any conceivable size.

Unfortunately, experience in using Eq. (3.46) is limited so all of its limitations and pitfalls may not be known. The following precautions seem advisable:

(a) Avoid using Eq. (3.46) for *d–s* combinations for which $f_d(s)$ is unmeasurably small. However, if it is small only because of a three-spin effect, it can probably be used without error.

(b) Internuclear distances determined between spins where a direct positive NOE enhancement is not observed will likely be inaccurate; avoid using such distances in further calculations.

(c) Calculations should be checked with the closed form approximations to get a feeling for the errors involved.

(d) A careful accounting of errors should be included in the computer analysis.

The above precautions are perhaps excessively conservative; only experience can be a realistic guide.

Finally, if the formulas of this chapter are used to estimate unmeasurable enhancements, these enhancements must be used cautiously as they contain systematic errors due to the approximations used.

K. Heteronuclear NOE

The theory for heteronuclear NOE between spins-$\tfrac{1}{2}$ is not much different from that of homonuclear NOE. However, the ratio of γ's which appeared in Eq. (3.6) indicates that: (a) unless the γ's are nearly equal only experiments where the spin of lowest γ is observed are practical; (b) when the spin of lowest γ is observed, enhancements may be found which are very large compared to those found in homonuclear cases. The second point is particularly interesting since spins of low γ typically have poor signal-to-noise ratios which can usefully be improved by the NOE. This was demonstrated by Kuhlmann and Grant (*17*) when they observed the $^{13}\text{C}\{^1\text{H}\}$ NOE in formic acid and obtained $f_\text{C}(\text{H}) = 1.98 \pm 0.15$ compared to the theoretical maximum ratio of 1.988. Similar results were obtained on adamantane by Kuhlmann *et al.* (*12*). In contrast, if the $^1\text{H}\{^{13}\text{C}\}$ enhancement could be measured it would have a maximum of 0.126 which would surely be reduced by intermolecular relaxation (i.e., ρ^*).

Centrally located spins, such as ^{13}C, will often have reduced ρ^* values because they are protected from intermolecular relaxation by their surrounding protons. However such spins may have other mechanisms such as spin-rotation contributing to ρ^* which will reduce the NOE enhancement. This was demonstrated for ^{13}C in methyl iodide by Gillen *et al.* (*18*).

It is surprising that heteronuclear NOE has not received wider attention. The maximum NOE enhancements ($\rho^* = 0$) for some experiments are listed on Table 3-8.

TABLE 3-8

MAXIMUM NOE ENHANCEMENTS (ρ^* ASSUMED ZERO FOR THE OBSERVED NUCLEUS) FOR VARIOUS HETERONUCLEAR NOE EXPERIMENTS

Observe	$\{^1\text{H}\}$	$\{^{19}\text{F}\}$	$\{^2\text{D}\}$
^1H	0.50	0.47	0.077
^{13}C	1.99	1.87	0.30
^{15}N	−4.93	−4.64	−0.76
^{19}F	0.53	0.50	0.082
^{29}Si	−2.52	−2.37	−0.39
^{31}P	1.24	1.16	0.19

With respect to the negative enhancements it should be noted that $f = -1$ corresponds to no signal at all so, e.g., if the maximum enhancement for $^{15}\text{N}\{^1\text{H}\}$ is achieved, $f = -4.93$ but the actual magnitude of the ^{15}N NMR intensity is larger by only 3.93. (Signal-to-noise improvements, of course, may be better than 3.93 if fine structure due to J coupling is collapsed by the double resonance.)

It may appear in Eq. (3.32) that the ratio $(\gamma_x/\gamma_m)^3$ would give an advantage if $\gamma_m \ll \gamma_x$. Such is not usually the case. Let us examine a specific case with $^{13}\text{CH}_2$. If one of the protons is in common in Eq. (3.32), the right-hand side is multiplied by $(\gamma_\text{H}/\gamma_\text{C})^3 \simeq 64$. However, in this case

$$\left(\frac{r_{\text{H}1,\text{H}2}}{r_{\text{H}1,\text{C}}}\right)^6 = (64)\frac{f_{\text{H}1}(\text{C}) + f_{\text{H}1}(\text{H}2)f_{\text{H}2}(\text{C})}{f_{\text{H}1}(\text{H}2) + f_{\text{H}1}(\text{C})f_\text{C}(\text{H}2)}$$

It can be seen that the numerator is small—and perhaps unmeasurable—and the small $f_H(C)$ enhancements are not made more accurate by multiplying by 64.

On the other hand, if the ^{13}C nucleus is in common the ratio

$$\left(\frac{r_{C,H1}}{r_{C,H2}}\right)^6 = \frac{f_C(H2) + f_C(H1)\,f_{H1}(H2)}{f_C(H1) + f_C(H2)\,f_{H2}(H1)}$$

is multiplied by $\gamma_H/\gamma_H = 1$ and can be determined quite accurately. There is still a problem in that the H{H} enhancements must be measured for the protons which are interacting with the ^{13}C; these signals may be obscured by the $^{12}C–H$ proton resonance.

L. Conclusion

Quantitative measurements of NOE enhancements can be useful in determining relative internuclear distances in spin systems with more than three spins. In such determinations the key experiments in locating a given spin are those in which that spin is observed while the others are saturated. Distances between spins for which a significant and direct positive NOE enhancement is not observed cannot be determined accurately. (If the gyromagnetic ratios of the spins have opposite sign, "direct" NOE enhancements will, of course, be negative.) The accuracy of NOE structural studies varies as much with geometry as with the distances involved and much remains to be learned about the limitations of the method. As a method for determining conformation in solution the NOE retains several formidable advantages: the theory is direct and simple, and only nuclei with spin enter into the problem.

Analysis of NOE data can be further simplified by several methods.

(a) Protons whose location is unimportant and which contribute significantly to the relaxation of the spins of interest can be replaced with deuterium—methyl groups are frequently in this class.

(b) "Linear" multispin systems can be approximated as a succession of three-spin systems.

(c) The effect of intramolecular relaxation of nuclei outside the subgroup whose geometry is sought can often be lumped with ρ^*; this is particularly useful when only one of the nuclei of the subgroup is enhanced significantly (positively *or* negatively) when the "outside" nuclei are saturated.

NOE in conformationally mobile and chemical exchanging systems, specifically excluded from this chapter, will be considered in Chapters 4 and 7, respectively.

REFERENCES

1. P. S. Hubbard, *Phys. Rev.* **109**, 1153 (1958); erratum **111**, 1746 (1958).
2. P. S. Hubbard, *Phys. Rev.* **128**, 650 (1962).
3. L. K. Runnels, *Phys. Rev. A* **134**, 28 (1964).
4. H. Schneider, *Ann. Phys.* [13], **7**, 313 (1964).
5. G. W. Kattawar and M. Eisner, *Phys. Rev.* **126**, 1054 (1962).
6. H. Schneider, *Z. Naturforsch. A* **19**, 510 (1964).
7. R. L. Hilt and P. S. Hubbard, *Phys. Rev. A* **134**, 392 (1964).
8. H. Schneider, *Ann. Phys.* [16], **7**, 135 (1965).
9. J. Noggle, *J. Phys. Chem.* **72**, 1324 (1968).
10. P. S. Hubbard, *J. Chem. Phys.* **51**, 1647 (1969).
11. P. S. Hubbard, *J. Chem. Phys.* **52**, 563 (1970).
12. K. F. Kuhlmann, D. M. Grant, and R. K. Harris, *J. Chem. Phys.* **52**, 3439 (1970).
13. R. E. Schirmer, J. H. Noggle, J. P. Davis, and P. A. Hart, *J. Amer. Chem. Soc.* **92**, 3266 (1970); erratum; **92**, 7239 (1970).
14. R. A. Bell and J. K. Saunders, *Can. J. Chem.* **48**, 1114 (1970).
15. R. A. Bell and J. K. Saunders, *Can. J. Chem.* **46**, 3421 (1968).
16. J. P. Davis, unpublished results, 1970.
17. K. F. Kuhlmann and D. M. Grant, *J. Amer. Chem. Soc.* **90**, 7355 (1968).
18. K. T. Gillen, M. Schwartz, and J. H. Noggle, *Mol. Phys.* **20**, 899 (1971).

CHAPTER

4

THE EFFECTS OF INTERNAL MOTIONS

In previous chapters it was assumed that the molecules being studied were rigid, so that all internuclear distances were independent of time. In this chapter the dependence of the NOE on molecular conformation will be examined in more detail, and the discussion will be generalized to include the effects of internal motions. We shall find that a general treatment is possible for the special case of internal motions that do not affect internuclear distances, but that the problem of more general motions has not been completely solved; the discussion of these general motions will be restricted to cases in which the motion is slow compared with τ_c.

Discussion of the class of internal motions referred to as intramolecular exchange will be deferred to Chapter 7. Properly speaking, chemical exchange implies a process in which chemical bonds are repeatedly made and broken, whereas an internal motion involves changes in molecular geometry without bond rupture. However, when an internal motion results in the transfer of a spin or group of spins between a few distinct positions and each position gives rise to a separate NMR signal, then the internal motion can be treated exactly the same as a problem in chemical exchange between these positions. Internal motions satisfying this condition will be referred to as *intramolecular exchange* and will be discussed in Chapter 7. That this separation of intramolecular exchange from other internal motions is rather arbitrary can be seen by

76

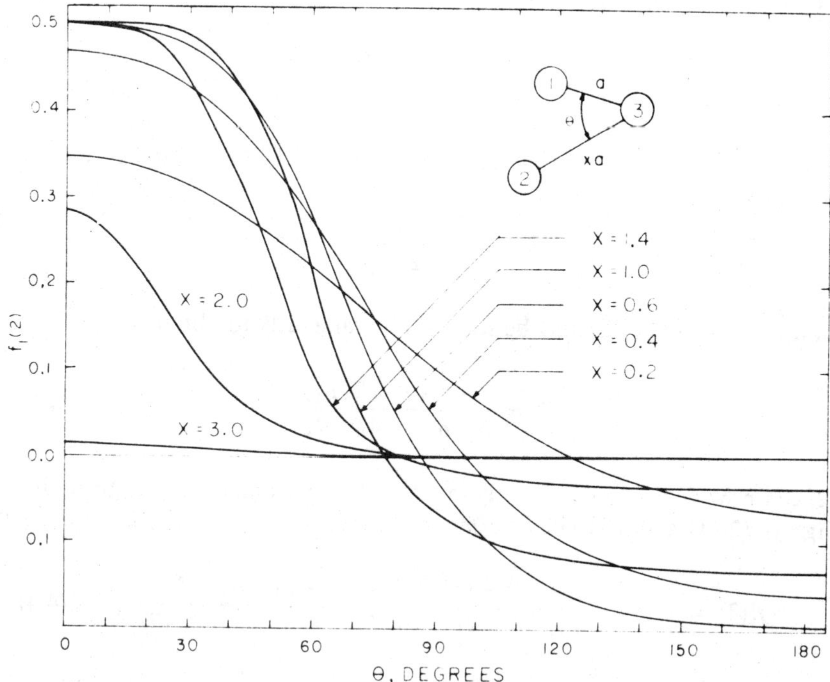

Fig. 4-1   The dependence of the nuclear Overhauser effect on the relative positions of the spins in a three-spin system. The curves have been calculated using Eq. (4.4), which assumes $\rho^* = 0$.

considering the fact that when the rate of an intramolecular exchange process exceeds the frequency separation of the various sites in the NMR spectrum, the signals from all sites coalesce into a single NMR line, and the process is then treated as an internal motion.

A. Calculation of Enhancements as a Function of Internal Coordinates

We shall consider first the way in which the NOE in a group of three spins depends upon their spatial disposition. The effects of interactions with any other spins that might be present in the molecule will be ignored, as will the effects of intermolecular relaxation. The geometry of this group of three spins can be specified by the angle θ and the distances a and xa, as shown in Fig. 4-1. The formula for the enhancement of the resonance of spin 1 when spin 2 is irradiated was given in Chapter 3, Section G as

$$f_1(2) = \frac{\rho_{12} - \rho_{13} f_3(2)}{2R_1} \tag{4.1}$$

The analogous equation for the experiment in which 2 is irradiated and 3 observed is

$$f_3(2) = \frac{\rho_{32} - \rho_{31} f_1(2)}{2R_3} \tag{4.2}$$

Equations (4.1) and (4.2) may be solved simultaneously to obtain

$$f_1(2) = \frac{\rho_{13}\rho_{23} - 2R_3\rho_{12}}{\rho_{13}\rho_{31} - 4R_1 R_3} \tag{4.3}$$

$f_1(2)$ can now be expressed directly in terms of the geometry of the group by using Eq. (2.11) and a little trigonometry to give

$$f_1(2) = \frac{x^{-6} - 2(1 + x^{-6})(1 + x^2 - 2x\cos\theta)^{-3}}{1 - 4(1 + x^{-6}) - 4(1 + x^{-6})(1 + x^2 - 2x\cos\theta)^{-3}} \tag{4.4}$$

Plots of $f_1(2)$ versus θ calculated using Eq. (4.4) are shown in Fig. 4-1 for several values of x. The following general observations can be derived from the figure:

(1) When θ is small and x is close to 1, spins 1 and 2 are close to one another and the enhancement between them is large and positive.

(2) As θ increases toward 180°, spins 1 and 2 become closer to 3 than to one another, and the enhancements become negative.

(3) As x increases beyond 1, the distance between spins 1 and 2 becomes greater than that between 1 and 3, so the magnitude of $f_1(2)$ rapidly drops off for all values of θ; by the time $x = 3$, only a very small positive enhancement for $\theta \approx 0$ is found.

(4) Regardless of the value of x, there exists an angle for which $f_1(2) = 0$, and the distance between spins 1 and 2 is not necessarily large at this angle (e.g., $x = 0.40$ at $\theta = 97°$).

These observations emphasize the fact that it is the relative disposition of all interacting spins that determines the NOE rather than simply the distance between the irradiated and observed spin.

The enhancements discussed in the preceding paragraph were easily calculated using the equations developed in Chapter 3. However, increasing the number of spins present results in a very rapid increase in the complexity of the closed-form equations, so that an alternative, more convenient computational procedure is desirable for multispin cases. We will discuss a suitable procedure and then apply it to calculating the enhancements as a function of internal angle for calculating the enhancements as a function of internal angle for angle for 2′,3′-isopropylideneinosine (Fig. 4-2).

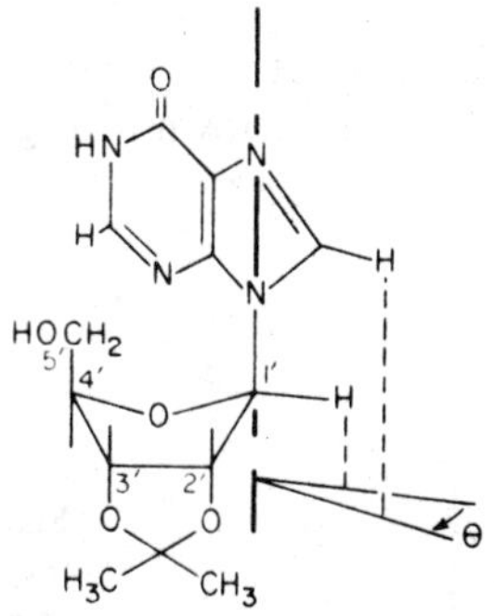

Fig. 4-2 2′,3′-Isopropylideneinosine. The angle θ describes the orientation of the base relative to the ribose moiety.

A general expression for the enhancement of the resonance of spin d when the resonances of spins s are saturated was given by Eq. (3.6) as

$$f_d(s) = \sum_s \gamma_s \rho_{ds}/2\gamma_d R_d - \sum_n \gamma_n \rho_{dn} f_n(s)/2\gamma_d R_d \tag{4.5}$$

The sum over s includes all spins whose resonances are saturated and the sum over n includes all spins except d and s. The computational procedure is based upon the observation that the first term in Eq. (4.5) represents the direct interaction of the observed and irradiated spins and generally will be of greater magnitude than the terms in the second sum which represent indirect interactions mediated by a third spin ($s \rightarrow n \rightarrow d$). Thus all the enhancements may be estimated by using just the first term of (4.5), or

$$f_n(s) \approx \sum_s \gamma_s \rho_{ns}/2\gamma_n R_n \tag{4.6}$$

These estimates are then improved iteratively. The complete procedure is:

(1) Substitute the internuclear distances corresponding to the conformation of interest into Eqs. (3.2) and (3.3) to obtain values for all the ρ's and R's.
(2) Using Eq. (4.6), estimate all of the $f_n(s)$, including $f_d(s)$.
(3) Substitute the values of ρ_{ns}, R_n, and the estimated values of the $f_n(s)$ back into Eq. (4.5) to obtain new estimates of all the $f_n(s)$.

(4) The values obtained for the $f_n(s)$ are improved iteratively by repeating step (3) as often as necessary. Iteration is terminated when all $f_n(s)$ obtained on the $(n+1)$th iteration agree to within a preset limit with the values obtained on the nth iteration.

Let $f_d(s, \Omega)$ denote the enhancement of the resonance of d when s is saturated and the molecule is in conformation Ω. The input to a computer program to calculate $f_d(s, \Omega)$ would consist of values of ρ^* and a set of geometric parameters from which the internuclear distances could be calculated within the program. A suitable set of geometric parameters for the case of a single internal axis of rotation are shown in Fig. 4-3. A particular conformation of the molecule is

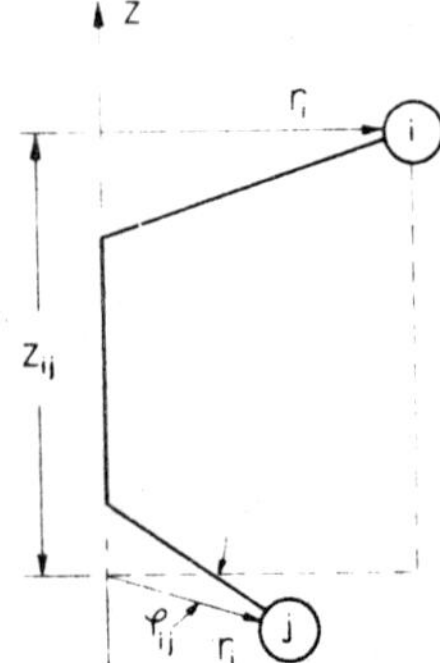

Fig. 4-3 Definition of geometric parameters used to describe the relative positions of the spins in the presence of a single internal rotation.

selected as a reference conformation. The relative positions of spins in the molecule are then characterized by their perpendicular distances r_i from the axis of internal rotation, their separation along the axis of rotation z_{ij} and the phase angle ϕ_{ij} between the projections of r_i and r_j on a plane perpendicular to the z axis. If the molecule is rotated through an angle θ from the reference conformation, the internuclear distances are given by

$$r_{ij}(\theta) = [z_{ij}^2 + r_i^2 + r_j^2 - 2r_i r_j \cos(\phi_{ij} + \theta)]^{\frac{1}{2}} \qquad (4.7)$$

Provision must be made in the calculation so that internuclear distances which do not depend on θ are treated as constants. In more complicated cases of internal motions, more sophisticated methods of calculating internuclear distances, such as that of Hilderbrandt (1), might prove more useful.

As an example of the NOE's expected in more complicated molecules, the iterative procedure has been applied† to calculate the enhancements as a

---

† The calculations of enhancements in 2', 3'-isopropylideneinosine were performed by P. A. Hart and J. P. Davis in collaboration with the authors.

function of θ in 2′,3′-isopropylideneinosine (Fig. 4-2). In order to obtain realistic values for the enhancements in this molecule, the geometric parameters required in the calculation (see Fig. 4-3) were measured from a model constructed with a Framework Molecular Models Set (Prentice-Hall, Inc.). The model was constructed with the glycosidic bond (C1′–N9) lying in the plane of the purine base. The small angles found between the glycosidic bond and the plane of the base in X-ray crystal studies of other purine nucleosides [less than 8° (2)] would not cause a significant change in internuclear distances and thus would not affect the magnitude of the NOE observed in these molecules. The ribose conformation selected for these calculations had an H1′–H2′ dihedral angle of approximately 99°, and an H2′–H3′ dihedral angle of about 8° (Case II of Table 4-1).

TABLE 4-1

DEPENDENCE OF THE EXTREME VALUES OF $f_i(j,\theta)$ ON RIBOSE CONFORMATION AND INTERMOLECULAR RELAXATION

		Case I		Case II		Case III		Case IV	
Saturate	Observe	Angle	$f_j(k)_{max}$	Angle	$f_j(k)_{max}$	Angle	$f_j(k)_{max}$	Angle	$f_j(k)_{max}$
1′	8	339	0.48	349	0.46	339	0.46	0	0.32
2′	8	117	0.48	106	0.46	117	0.49	106	0.44
3′	8	180	0.04	169	0.21	191	0.26	169	0.17
1′	2[a]	286	−0.04	275	−0.04	286	−0.04	286	−0.03
2′	2	286	0.46	286	0.48	286	0.46	286	0.41
3′	2	339	0.12	339	0.27	0	0.35	339	0.16
8	1′	0	0.29	0	0.21	0	0.23	0	0.18
8	2′	117	0.41	117	0.35	117	0.45	117	0.34
4′	1′	191	0.04	127	0.05	191	0.13	206	0.04
H1′–H2′ Dihedral angle		91°		99°		103°		99°	
H2′–H3′ Dihedral angle		8°		9°		21°		9°	

[a] The experimental value for $f_2(1′)$ is known to be negative, so the negative extremum and its location are recorded in the table.

The two 5′ protons present a problem because nothing is known of their motion about the C5′–C4′ bond. A very approximate treatment of the interactions of 5′ was considered adequate providing values for $f_n(5′)$, $f_{5′}(n)$, and $f_{4′}(n)$ were not of interest. The enhancements involving H4′ are excluded because the 5′ protons probably contribute heavily to the relaxation of H4′: thus an inexact treatment of the H4′–H5′ interaction would result in a large error in the calculation of $f_{4'}(n)$. The approximation used was to set

$$r_{n5'}^{-6} = \tfrac{1}{2}(r_{n5'\,\text{min}}^{-6} + r_{n5'\,\text{max}}^{-6})$$

and to ignore any difference in the correlation time for the interactions of the 5' protons as compared with the correlation times for other spin interactions in the molecule. The isopropylidene protons were ignored entirely in these calculations.

As shown in Fig. 4-2, the orientation of the base with respect to the ribose was specified by the angle θ between the plane of the base and the plane defined by H1', C1', and N9. Viewing along the glycosidic bond from C1' toward N9, θ is taken as positive when the base is rotated counterclockwise with respect to the ribose and is zero when H8 lies above H1'. The sense of this rotation is the same as that in the convention set by Donohue and Trueblood (3) for nucleosides and nucleotides, but our 0° corresponds to their $\phi_{CN} = +120°$.

The enhancements that would result from this geometry were calculated as a function of θ using the iterative procedure discussed in the preceding section. Iteration was continued until the enhancements calculated on the nth iteration were within 3% of their values on the $(n-1)$th iteration. The calculated enhancements are plotted as a function of θ in Fig. 4-4. It can be seen that the NOE is very strongly dependent on molecular conformation, and that the angles at which maximum or minimum enhancements occur vary widely from experiment to experiment. If the molecule possessed a single, very strongly preferred conformation, a comparison of experimental with calculated enhancements would indicate the preferred value of θ. The ultimate limit in specifying this conformation on the basis of NOE experiments would be the breadth of the maxima and minima in the enhancement versus θ curves. There are only a few angles for which a small change in conformation would result in large changes in the enhancements. Looking at Fig. 4-4 we might estimate that, in the region around $\theta = 120°$, an uncertainty in an experimental enhancement of as little as 1% could result in an uncertainty of $\pm 20°$ in determining the conformation. Other conformations are more favorable because at least one of the experiments will prove to be quite sensitive to θ in most of them. In practice, the ultimate resolution in conformation is unlikely to be attained. The first reason for this is that there are rather large uncertainties in the calculated enhancement versus θ curves themselves. These arise because the basic geometry of the molecule in solution will not be known with any accuracy (e.g., the uncertain conformation of the isopropylideneribose in our example), and the magnitudes of the intermolecular relaxation will not always be known or easily estimated. These factors will be discussed in more detail later.

A second reason for reduced sensitivity to conformation is the fact that, in

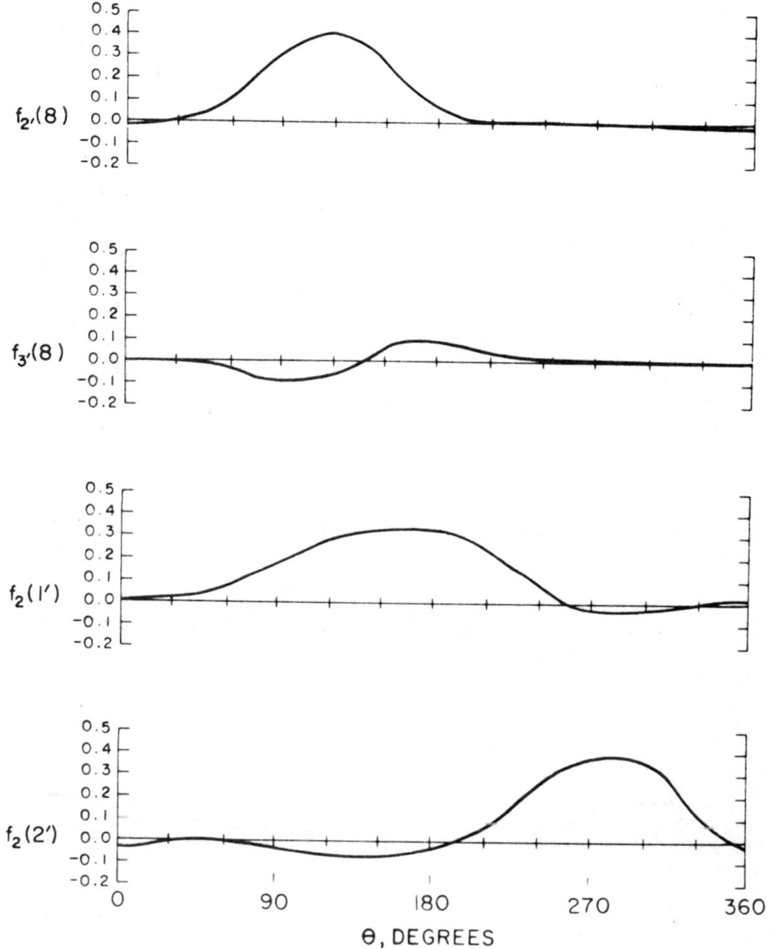

Fig. 4-4a   Legend on page 85.

general, there will not be a single conformation present but rather a distribution of conformations. The presence of a range of conformations introduces
additional approximations into the interpretation of the experimental data,
and thus some additional uncertainty in the results. These uncertainties do not
detract from the usefulness of the NOE method in conformational studies, but
do require that caution be used in arriving at a conclusion. The effects of
internal motions will be discussed in detail in Sections B and C.

To show the way in which the ribose conformation and intermolecular
relaxation affect the Overhauser experiments on 2′,3′-isopropylideneinosine,

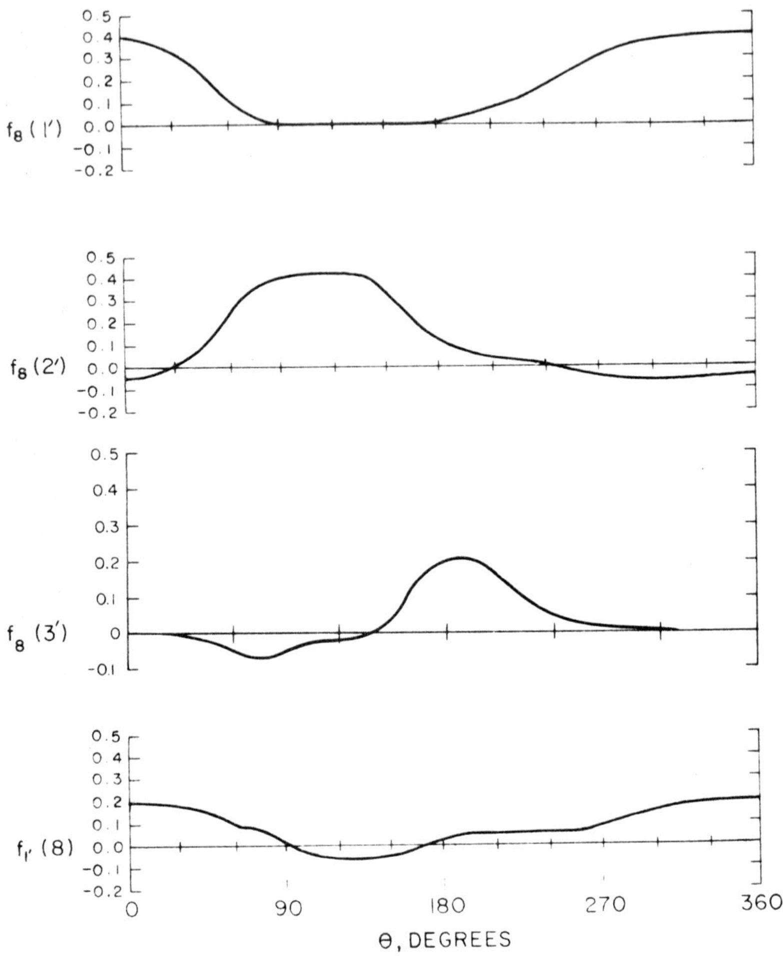

Fig. 4-4b

calculations were carried out for three different ribose conformations as well as one case where intermolecular relaxation was included. These are Cases I–IV of Table 4-1, respectively. The maximum values of the calculated enhancements are recorded in the table along with the angles at which the maxima occur.

The result of differing ribose conformations follows closely what one would intuitively expect from an examination of molecular models. Proceeding from Case I to Case III, the distance between C5′ and N9 increases and the molecule progresses to a more "open" structure. The way in which the distance between

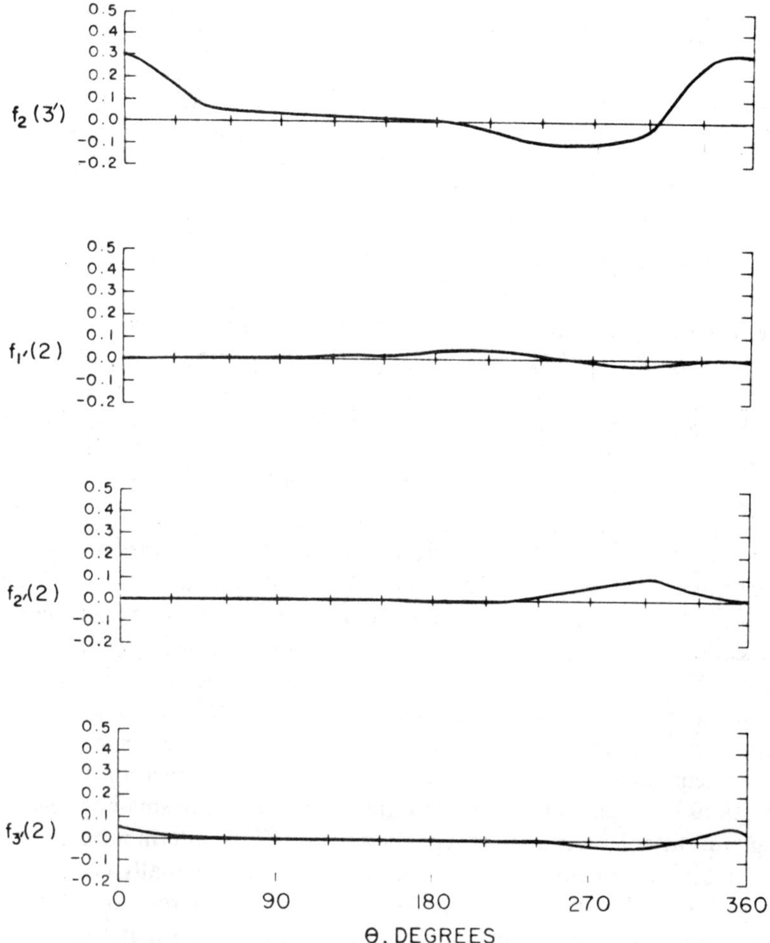

Fig. 4-4c   The nuclear Overhauser effects as a function of angle about the glycosidic bond in 2′,3′-isopropylideneinosine. The angle θ is defined in Fig. 4-2. The ribose geometry is that of Case II, Table 4-1, and it was assumed that $\rho^* = 0$ for all nuclei.

a given ribose proton and the base protons changes as the structure becomes more open depends upon the position of that proton in the ribose; the distance increases most rapidly at the 5′ end of the ribose and least rapidly at the 1′ end. Distances between ribose protons—particularly the H4′–H1′ distance—are also strongly affected. The effect of ribose conformation on the calculated enhancements follows the same pattern as its effect on internuclear distances. $f_8(1')$ and $f_8(2')$ are almost independent of ribose conformation, as are $f_2(1')$

and $f_2(2')$. On the other hand, $f_2(3')$ and $f_8(3')$ depend strongly on the ribose conformation, as does the intraribose enhancement $f_1(4')$. $f_1(8)$ and $f_2(8)$ are also somewhat dependent on the ribose conformation. The most important point arising from this discussion is that some of the enhancements observed in a molecule will depend almost exclusively upon one aspect of the conformation of the molecule, while others will depend upon all aspects of the conformation. Recognizing at the outset which variables will affect each enhancement should prove very helpful in deciding whether or not the NOE will be applicable to a particular problem in chemical structure, and if so, in designing the experiments and evaluating the results.

The effect of intermolecular relaxation on the observed enhancements is also shown in Table 4-1 as the $f_j(k)_{\max}$ for Case IV were computed using the same geometry as Case II, but including a contribution from intermolecular relaxation. The contribution was assumed to be the same for all spins and was given the value

$$\rho^* = h^2\gamma^4\tau_c a \times 10^{48} \tag{4.8}$$

with $a = 0.001$ Å^{-6}. This would correspond to a relaxation time (T_1) of 20 sec for all spins if there were no intramolecular relaxation and if $\tau_c = 10^{-10}$ sec.† The effect of intermolecular relaxation is to reduce the maximum enhancement observed in any given experiment, but the magnitude of the decrease is quite variable. For a proton such as H8 which does not interact strongly with more than one other spin, the intermolecular contribution may be an important part of R_{H8}, so relatively small increases in ρ^*_{H8} lead to large decreases in the enhancements expected. Other spins, such as H2′, have several rather strong interactions so small increases in ρ^* and thus in R result in small or negligible changes in the enhancements observed on them. As the intermolecular contribution increases, all enhancements will decrease and eventually become zero. This makes it very clear that sources of intermolecular relaxation must be minimized in order to maximize the information on conformation attainable from a set of NOE experiments.

B. Internal Motions Not Affecting Internuclear Distances

The presence of internal motion requires that ρ and σ be evaluated with the internal motion taken explicitly into account. In case the motion does not

---

† An order of magnitude estimate of the tumbling time for reasonably spherical or globular molecules in water at room temperature is obtained by taking the molecular weight in thousands and attaching the units nanoseconds to it. For other solvents the estimate may be made by multiplying the value in water by the ratio of the viscosity of the second solvent to that of water.

affect the internuclear distances (e.g., the $^{13}C-^{1}H$ distances in ethane) this problem can be solved quite generally. Woessner *et al.* (*4*) have carried out the necessary calculations for a molecule which approximates an axially symmetric ellipsoid with different rotational rate constants about the major and minor axes. There is an axis of internal rotation z' which makes an angle α with the major axis of the ellipsoid. The internuclear vector r forms an angle Δ with z', as shown in Fig. 4-5. While the motion of the ellipsoid was assumed to

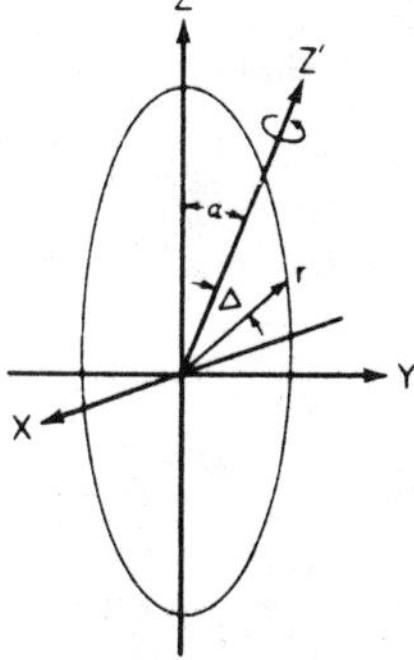

Fig. 4-5 The coordinates used to calculate the effective correlation time τ_e for spins in an anisotropically tumbling ellipsoid in the presence of internal motions. [Woessner *et al.* (*4*).]

be rotational diffusion, two models were considered for the internal motion. The first model was that of random jumping between three equivalent positions, as might be the case if the spins concerned were the protons of a methyl group. The second model was a rotational diffusion model in which all angles about z' were assumed to be equally probable. In the extreme narrowing region the result is that Eqs. (2.10) and (2.11) remain valid provided τ_c is replaced by the effective correlation time τ_e, given in Table 4-2, and we still have $\sigma_{ij} = S_i(S_i+1)\,\rho_{ij}/2S_j(S_j+1)$. Calculations for other models of the motion can be found in the literature (*5–8*) and lead to the same conclusions.

The expressions in Table 4-2 show that the only way in which the rate constant ($\mathscr{R}$ or $\mathscr{D}$) for the internal motion enters into the problem is through various sums with the rate constants for molecular tumbling ($\mathscr{R}_1$ and $\mathscr{R}_2$). Thus if $\mathscr{R}$ (or $\mathscr{D}$) is much smaller than $\mathscr{R}_1$ and $\mathscr{R}_2$ it will have no effect on τ_e and the internal motion will not affect the magnitude of the NOE observed on the molecule. However, when the rate of the internal motion is on the same order of magnitude as $\mathscr{R}_1$ and $\mathscr{R}_2$, the effective correlation time for the interaction of i and j will decrease and this will result in a decrease in ρ_{ij}. As i and j will generally have other relaxation paths available to them which are unaffected by the internal motion, ρ_{ij}/R_i and σ_{ij}/R_j will also decrease and a smaller enhancement will result.

TABLE 4-2

THE EFFECTIVE CORRELATION TIME IN A MOLECULE WITH INTERNAL ROTATION[a]

Case I. Random Jumping Between 0°, 120°, and 240°

$$\tau_e = B_{A1}\tau_A + (B_{A2} + B_{A3})\tau_{A2} + B_{B1}\tau_B + (B_{B2} + B_{B3})\tau_{B2} + B_{C1}\tau_C + (B_{C2} + B_{C3})\tau_{C2}$$

Case II. Rotational Diffusion with All Angles Equally Probable

$$\tau_e = B_{A1}\tau_A + B_{A2}\tau_{A2'} + B_{A3}\tau_{A3'} + B_{B1}\tau_B + B_{B2}\tau_{B2'} + B_{B3}\tau_{B3'} + B_{C1}\tau_C + B_{C2}\tau_{C2'} + B_{C3}\tau_{C3'}$$

Definitions of Constants:

$$B_{A1} = \tfrac{1}{8}(1 - 3\cos^2\alpha)^2(1 - 3\cos^2\Delta)^2 \qquad B_{B1} = \tfrac{3}{8}\sin^2 2\alpha(3\cos^2\Delta - 1)^2$$
$$B_{A2} = \tfrac{9}{16}\sin^2 2\alpha \sin^2 2\Delta \qquad B_{B2} = \tfrac{3}{4}(\cos^2 2\alpha + \cos^2\alpha)\sin^2 2\Delta$$
$$B_{A3} = \tfrac{9}{16}\sin^4\alpha \sin^4\Delta \qquad B_{B3} = \tfrac{3}{4}(\sin^2\alpha + \tfrac{1}{4}\sin^2 2\alpha)\sin^4\Delta$$

$$B_{C1} = \tfrac{3}{8}\sin^4\alpha(3\cos^2\Delta - 1)^2$$
$$B_{C2} = \tfrac{3}{4}(\sin^2\alpha + \tfrac{1}{4}\sin^2 2\alpha)\sin^2 2\Delta$$
$$B_{C3} = \tfrac{3}{16}[(1 + \cos^2\alpha)^2 + 4\cos^2\alpha]\sin^4\Delta$$

$$1/\tau_A = 6\mathcal{R}_2 \qquad 1/\tau_B = \mathcal{R}_1 + 5\mathcal{R}_2 \qquad 1/\tau_C = 4\mathcal{R}_1 + 2\mathcal{R}_2$$
$$1/\tau_{A2} = 6\mathcal{R}_2 + \mathcal{R} \qquad 1/\tau_{B2} = \mathcal{R}_1 + 5\mathcal{R}_2 + \mathcal{R} \qquad 1/\tau_{C2} = 4\mathcal{R}_1 + 2\mathcal{R}_2 + \mathcal{R}$$
$$1/\tau_{A2'} = 6\mathcal{R}_2 + \mathcal{D} \qquad 1/\tau_{B2'} = \mathcal{R}_1 + 5\mathcal{R}_2 + \mathcal{D} \qquad 1/\tau_{C2'} = 4\mathcal{R}_1 + 2\mathcal{R}_2 + \mathcal{D}$$
$$1/\tau_{A3'} = 6\mathcal{R}_2 + 4\mathcal{D} \qquad 1/\tau_{B3'} = \mathcal{R}_1 + 5\mathcal{R}_2 + 4\mathcal{D} \qquad 1/\tau_{C3'} = 4\mathcal{R}_1 + 2\mathcal{R}_2 + 4\mathcal{D}$$

$\mathcal{R}_1$ = rotational diffusion constant of the ellipsoid about its major axis

$\mathcal{R}_2$ = rotational diffusion constant of the ellipsoid about its minor axis

$\mathcal{R}$ = three-halves the total rate with which the spin jumps from any of its three equivalent positions

$\mathcal{D}$ = rotational diffusion constant for the motion about the internal axis

[a] The molecule is treated as an ellipsoid undergoing anisotropic rotational diffusion. The angles used in the table are defined in Fig. 4-5. The effective correlation times are given in the extreme narrowing limit although the original work (4) was not restricted to this case.

C. The Effects of Slow Internal Motions

Let us consider an internal motion with a rate constant $k \ll \tau_c^{-1}$, where τ_c is the correlation time for molecular reorientation. A motion satisfying this condition will not contribute to the effective correlation time for the interaction of any spin pairs in the molecule and so the motion may be treated using the adiabatic equations of McConnell (9). The path of the observed spin d with respect to other spins in the molecule is first divided into a large number of segments, N. McConnell's equation for the magnetization due to spins of type d which occur on the ith segment may then be written

$$d\langle \mathbf{I}_{zd}(i)\rangle/dt = -R_d(i)[\langle \mathbf{I}_{zd}(i)\rangle - I_{0d}(i)] - \sum_n \sigma_{dn}(i)[\langle \mathbf{I}_{zn}(i)\rangle - I_{0n}(i)]$$

$$+ k_{i+1,i}\langle \mathbf{I}_{zd}(i+1)\rangle + k_{i-1,i}\langle \mathbf{I}_{zd}(i-1)\rangle \qquad (4.9)$$

$$- (k_{i,i+1} + k_{i,i-1})\langle \mathbf{I}_{zd}(i)\rangle$$

where $R_d(i) = \sum_n \rho_{nd}(i) + \rho_d{}^*(i)$, $k_{i,i+1}$ is the rate constant for transfer of spins from the ith to the $(i+1)$th interval, and n runs over all spins other than d. The solution of (4.9) for $f_d(s)$ will be considered for two ranges of the values of k.

Case I. Very Slow Motion: $k \ll R_d$

Let s denote the spins whose resonances are saturated. Under steady-state conditions the derivative in (4.9) is zero and the equation may be rearranged to

$$\langle \mathbf{I}_{zd}(i)\rangle - I_{0d} = \sum_s \frac{\sigma_{ds}(i) I_{0s}(i)}{R_d(i)} - \sum_{n \neq d,s} \frac{\sigma_{dn}(i)}{R_d(i)} [\langle \mathbf{I}_{zn}(i)\rangle - I_{0n}(i)]$$

$$+ \frac{1}{R_d(i)} [k_{i+1,i}\langle \mathbf{I}_{zd}(i+1)\rangle + k_{i-1,i}\langle \mathbf{I}_{zd}(i-1)\rangle$$

$$- (k_{i,i+1} + k_{i,i-1})\langle \mathbf{I}_{zd}(i)\rangle]$$

By our assumption that $k_{ij} \ll R_d$ for all i and j, the last term drops out of the equation. Using $\sigma_{dn}(i) = I_d(I_d+1)\rho_{dn}(i)/2I_n(I_n+1)$, $I_{0n} \propto I_n(I_n+1)\gamma_n$ (see Chapter 3, Section C) and dividing through by $I_{0d}(i)$ we obtain

$$f_d(s,i) = \sum_s \frac{\gamma_s \rho_{ds}(i)}{2\gamma_d R_d(i)} - \frac{1}{2\gamma_d R_d(i)} \sum_{n \neq d,s} \gamma_n \rho_{dn}(i) f_n(s,i) \qquad (4.10)$$

The observed enhancement will then be given by

$$f_d(s) = \sum_i P_i f_d(s,i) \qquad (4.11)$$

where P_i is the fraction of the spins which occur on the segment i (equivalently, the fraction of the molecules in conformation i).

In the limit as $N \to \infty$ and the intervals become smaller and smaller, the sum in (4.11) is replaced by an integral. Equation (4.11) may then be generalized to

$$f_d(s) = \int f_d(s,\Omega) P(\Omega)\, d\Omega \qquad (4.12)$$

where Ω is the vector of internal coordinates describing the conformation, $P(\Omega)$ is the probability of conformation Ω, and the integral is over all possible conformations. The procedure used to calculate $f_d(s)$ is to first calculate a table of values of $f_d(s, \Omega)$ as a function of Ω. Since Eq. (4.10) (with i replaced everywhere by Ω) is identical in form to Eq. (4.5), the iterative procedure described in Section A may be used to calculate the $f_d(s, \Omega)$. Using the table of enhancements computed in this manner, the integral in (4.12) is evaluated numerically, perhaps using one of the procedures described in Appendix II.

Case II. Moderate Rates of Motion: $R_d \ll k \ll \tau_c^{-1}$

To obtain an expression for $f_d(s)$ when $R_d \ll k \ll \tau_c^{-1}$, we begin by setting the derivative in (4.9) equal to zero and then summing over i to obtain

$$\sum_i R_d(i)\left[\langle \mathbf{I}_{zd}(i)\rangle - I_{0d}(i)\right] + \sum_i \sum_n \sigma_{dn}(i)\left[\langle \mathbf{I}_{zn}(i)\rangle - I_{0n}(i)\right] = 0 \quad (4.13)$$

The k's have canceled identically because closure of the path requires $N + i = i$, and thus

$$\sum_i k_{i,i+1}\langle \mathbf{I}_{zd}(i)\rangle = \sum_i k_{i-1,i}\langle \mathbf{I}_{zd}(i-1)\rangle$$

$$\sum_i k_{i,i-1}\langle \mathbf{I}_{zd}(i)\rangle = \sum_i k_{i+1,i}\langle \mathbf{I}_{zd}(i+1)\rangle$$

In order to solve (4.13) for $f_d(s)$ we must obtain a relation between $\langle \mathbf{I}_{zn}(i)\rangle$, $\langle \mathbf{I}_{zn}\rangle$, and P_i. The required relation will be derived by first obtaining relations between $\langle \mathbf{I}_{zn}(i)\rangle$ and k_{ij} and between P_i and k_{ij}; the final step will then be to eliminate k_{ij} between these equations.

Under steady-state conditions with $k \gg R_d$, Eq. (4.9) becomes

$$k_{i+1,i}\langle \mathbf{I}_{zd}(i+1)\rangle + k_{i-1,i}\langle \mathbf{I}_{zd}(i-1)\rangle - (k_{i,i+1} + k_{i,i-1})\langle \mathbf{I}_{zd}(i)\rangle = 0 \tag{4.14}$$

This equation may be written more compactly as

$$\sum_i k_{ij}\langle \mathbf{I}_{zd}(i)\rangle = 0 \tag{4.15}$$

by using the definition

$$k_{ij} = \begin{cases} k_{ij}, & j = i \pm 1 \\ -(k_{i,i+1} + k_{i,i-1}), & j = i \\ 0, & j \neq i, i \pm 1 \end{cases}$$

We may further simplify the notation by writing (4.15) in matrix form as

$$\mathbf{KI} = 0 \tag{4.16}$$

where $[\mathbf{K}]_{ij} = k_{ji}$ and $[\mathbf{I}]_i = \langle I_{zd}(i) \rangle$. Equation (4.16) cannot be solved directly for the relation we want between $\langle I_{zn}(i) \rangle$ and the k_{ij} because the determinant of $\mathbf{K}$ is zero. This problem can be overcome by using the additional relation

$$\langle \mathbf{I}_{zd} \rangle = \sum_i \langle I_{zd}(i) \rangle \tag{4.17}$$

which can be written in matrix form as

$$\mathbf{I}_d = \mathbf{CI} \tag{4.18}$$

where

$$\mathbf{I}_d = \begin{pmatrix} \langle \mathbf{I}_{zd} \rangle \\ 0 \\ 0 \\ \vdots \\ 0 \end{pmatrix}, \qquad \mathbf{C} = \begin{pmatrix} 1 & 1 & 1 & & \cdots & & 1 \\ 0 & 0 & 0 & & \cdots & & 0 \\ 0 & 0 & 0 & & \cdots & & 0 \\ \vdots & & & & & & \vdots \\ 0 & 0 & 0 & 0 & 0 & 0 & 0 \end{pmatrix}$$

Adding (4.16) and (4.18) we find

$$(\mathbf{C} + \mathbf{K})\mathbf{I} = \mathbf{I}_d$$

Since the determinant of $(\mathbf{C} + \mathbf{K})$ is not zero, we can use Cramer's rule to obtain

$$\langle I_{zd}(i) \rangle = \frac{\text{cofactor}[\mathbf{C} + \mathbf{K}]_{1i}}{\det[\mathbf{C} + \mathbf{K}]} \langle \mathbf{I}_{zd} \rangle \tag{4.19}$$

The relation between P_i and the k_{ij} is derived in an entirely analogous fashion. Since the distribution of molecules among the different conformations will be the equilibrium distribution, the principle of microscopic reversibility allows us to write

$$- k_{i,i+1} P_i + k_{i+1,i} P_{i+1} = 0$$

and

$$- k_{i,i-1} P_i + k_{i-1,i} P_{i-1} = 0$$

Adding these equations together,

$$- (k_{i,i+1} + k_{i,i-1}) P_i + k_{i+1,i} P_{i+1} + k_{i-1,i} P_{i-1} = 0 \tag{4.20}$$

Comparing this with (4.14) we can immediately write

$$\mathbf{KP} = 0 \tag{4.21}$$

where $[\mathbf{P}]_i = P_i$. Following the procedure used for $\langle \mathbf{I}_{zd} \rangle$, we now write the normalization condition, $\sum_i P_i = 1$, as

$$1 = \mathbf{CP} \tag{4.22}$$

where 1 is the vector with 1 as its first element and all other elements zero. Adding (4.22) to (4.21) and solving

$$P_i = \frac{\text{cofactor}[\mathbf{C} + \mathbf{K}]_{1i}}{\det[\mathbf{C} + \mathbf{K}]} \tag{4.23}$$

Substituting (4.23) into (4.19) we finally arrive at the relation

$$\langle \mathbf{I}_{zd}(i) \rangle = P_i \langle \mathbf{I}_{zd} \rangle \tag{4.24}$$

Substituting (4.24) into (4.13)

$$\sum_i \left\{ R_d(i) P_i [\langle \mathbf{I}_{zd} \rangle - I_{0d}] + \sum_n P_i \sigma_{dn}(i) [\langle \mathbf{I}_{zn} \rangle - I_{0n}] \right\} = 0 \tag{4.25}$$

Defining $\langle R_d \rangle = \sum_i R_d(i) P_i$, $\sigma_{dn} = \sum_i P_i \sigma_{dn}(i)$, and carrying the summation over i through Eq. (4.25) we find

$$\langle R_d \rangle [\langle \mathbf{I}_{zd} \rangle - I_{0d}] = -\sum_n \langle \sigma_{dn} \rangle [\langle \mathbf{I}_{zn} \rangle - I_{0n}]$$

The enhancement of the resonance of spin d when spins s are saturated is then

$$f_d(s) = \frac{1}{2} \sum_s \frac{\langle \rho_{ds} \rangle}{\langle R_d \rangle} - \frac{1}{2} \sum_s \frac{\langle \rho_{dn} \rangle}{\langle R_d \rangle} f_n(s) \tag{4.26}$$

Thus we find that in the intermediate rate of exchange (R or $T_1^{-1} \ll k \ll \tau_c^{-1}$) it is the ρ_{ij} which must be averaged over the conformations present, rather than the $f_i(j)$ themselves. We found earlier [Eqs. (4.10) and (4.11)] that it is the $f_i(j)$ which must be averaged when $k \ll T_1^{-1}$. The difference arises because when $k \ll T_1^{-1}$, the molecule remains in each conformation long enough to come to steady state; the observed enhancement is then the ensemble average of the steady-state enhancements, $f_i(j, \Omega)$, as indicated in Eq. (4.12). However, when $k \gg T_1^{-1}$, the residence time in a given conformation is much too short for a steady state to be reached; in fact, all conformations will be sampled in the time it takes the system to achieve a steady state. The enhancements in this case must be calculated using the average of the interactions experienced by the spin, as indicated in Eq. (4.26).

D. The Problem of Interpreting NOE Experiments in the Presence of Internal Motion

The effects of internal motions on the NOE have been considered in some detail in the preceding sections. In this section we will take a brief look at the problem of interpreting the results of NOE experiments when such motions are present. Since work in this area is all very recent and incomplete, our discussion will be rather speculative in nature and the reader should not be surprised when he finds most of his questions still unanswered after reading this section.

First let us review what is known and what we would like to determine. We know the primary structure of the molecule including good estimates of bond lengths, bond angles, and probably many of the dihedral angles defining the conformation of the molecule. We also know the experimental enhancements, and furthermore can assume the solution being studied to be sufficiently dilute and of such a composition that significant pair formation or intermolecular NOE's can be dismissed.

We would like to know the main features of the function describing the distribution of the molecules among the possible conformations. In order to obtain this information it may be necessary to estimate the order of magnitude of ρ^* and of the rate constant for the internal motion as well.

The case discussed in Section B where the internuclear distances are independent of the internal rotation is the simplest of the three cases we have presented because the only effect of the internal motion is to introduce a second correlation time into the equations. In many cases of interest, the two correlation times could be determined independently by measuring the relaxation times of ^{13}C atoms in various sections of the molecule.

In the limit of very slow exchange ($k \ll R_d$) the procedure used might be to select a form for the distribution function, compute a table of values of $f_i(j, \Omega)$, and then perform a least-squares fit (see Appendix II) of the enhancements calculated using Eq. (4.12) to the experimental enhancements. How reliable such a result would be remains an open question. Two of the most obvious difficulties in this approach are (1) deciding whether or not the system being studied really satisfies the condition $k \ll R_d$, and (2) in establishing a reasonable description of the conformation and geometry of the molecule without increasing the number of parameters beyond the limit set by the number of experimental enhancements available.

The procedure described for very slow motions is equally applicable to the moderate rate region, $R_d \ll k \ll \tau_1^{-1}$. The only significant difference is that the distribution function selected is used with a table of values of r_{ij}^{-6} to calculate $\langle \rho_{ij} \rangle$, followed by application of Eq. (4.26), instead of using it with a table of the $f_i(j, \Omega)$ to compute $f_i(j)$. The major difficulties expected in treating a problem in this rate region are the same as in the slow region.

In the moderate rate region there is the possibility of an alternative approach to the problem. If a sufficient number of enhancements can be measured experimentally, it may prove advantageous to solve Eqs. (4.26) simultaneously for the $\langle \rho \rangle$'s first and then fit the distribution function to the $\langle \rho \rangle$ values rather than the $f_i(j)$ values. This is closely analogous to the procedure used in Chapter 3 in determining internuclear distances in rigid molecules. It must be kept in mind, however, that $\langle \rho_{ij} \rangle$ is proportional to $\langle 1/r_{ij}^6 \rangle$ and thus $\langle \rho_{ij} \rangle^{1/6} \propto \langle 1/r_{ij}^6 \rangle^{1/6}$ which does not have as simple an interpretation as it did in the rigid case where $\rho_{ij}^{1/6} = 1/r_{ij}$. It is, however, unlikely that this alternative approach to the problem will prove useful in very many cases because (1) it does require a more complete set of enhancements in order to be able to solve Eqs. (4.26) simultaneously for $\langle \rho_{ij} \rangle$; and (2) it is not easily decided which of the $\langle \rho_{ij} \rangle$ are actually negligible, as is done in rigid molecules. On the other hand, an advantage of this method is that no assumptions need to be made concerning the magnitude of the ρ^*'s.

In many situations in the moderate exchange region it may be informative to use the three- and four-spin formulas [Eqs. (3.32) and (3.41)] to estimate $\langle \rho \rangle$, since these formulas can simplify considerably if certain f's are near zero. If the distribution function has a single maximum and is rather narrow, the distinction between $\langle r \rangle$ and $\langle r^{-6} \rangle^{-1/6}$ may not be so large as to make the numbers estimated in this fashion entirely meaningless.

E. Conclusion

We have shown how to calculate the Overhauser effect expected in conformationally mobile systems, and have discussed briefly the possibility of obtaining approximate distributions of the molecules among the conformations by use of NOE data. Because of a general lack of experience with this method, its real potential in conformational problems cannot be evaluated at this time.

REFERENCES

1. R. L. Hilderbrandt, *J. Chem. Phys.* **51**, 1654 (1969).
2. M. Sundaralingam and L. H. Jensen, *J. Mol. Biol.* **13**, 930 (1965).
3. J. Donohue and K. N. Trueblood, *J. Mol. Biol.* **2**, 363 (1960).
4. D. E. Woessner, B. S. Snowden, Jr., and G. H. Meyer, *J. Chem. Phys.* **50**, 719 (1969).
5. D. E. Woessner, *J. Chem. Phys.* **36**, 1 (1962).
6. D. E. Woessner, *J. Chem. Phys.* **42**, 1855 (1965).
7. D. E. Woessner, *J. Chem. Phys.* **37**, 647 (1962).
8. W. T. Huntress, *J. Chem. Phys.* **48**, 3524 (1968).
9. H. M. McConnell, *J. Chem. Phys.* **28**, 430 (1958).

EXPERIMENTAL METHODS

A. Sample Preparation

In order to maximize NOE enhancements and minimize intramolecular relaxation effects, great care should be taken in preparing the NMR sample. The usual precautions applicable to the preparation of high-resolution NMR samples apply (*1a,b*) with greater than usual care taken to avoid turbid or viscous samples and to exclude paramagnetic impurities.

When possible, samples should be carefully degassed by freezing in a good vacuum (about 10^{-4} Torr or less); after freezing and pumping, the sample should be thawed in vacuum to permit entrained oxygen to escape, then frozen and pumped on again. Several of these freeze–pump–thaw cycles should be used. If no harm to the analytical sample will result, liquid nitrogen should be used for freezing. If the analytical sample would be harmed by freezing, bubbling gaseous nitrogen through the solution and sealing under nitrogen is probably adequate.

The compound whose NOE is being investigated should be at the minimum concentration consistent with an adequate signal-to-noise ratio. Since intensities must be measured carefully, the signal-to-noise problem is, if anything, more severe than in ordinary NMR; on the other hand, since field lock operation is routine, signal averaging represents no great additional experimental difficulty and should always be used when available. In any case concentrations should be less than 10% (v/v) and preferably less than 5%. Definitive studies on the effect of solute concentrations have not been made. Obviously, the compound being studied should be as pure as possible, preferably with superfluous protons replaced by deuterium to the greatest feasible extent.

Nonmagnetic or low-magnetic solvents are absolutely necessary. Solvents such as CS_2, CCl_4, OCS, or SO_2 are preferred, but deuterated organic solvents are acceptable. Deuterated solvents with 1% proton impurities are adequate providing the proton impurity resonance does not interfere with the NOE experiments. If solvents of higher purity are available, they should be used.

NOE experiments must be carried out under magnetic field lock conditions (see p. 101). Internal proton or fluorine lock should be avoided. Some spectrometers permit use of an external lock sample or an internal lock on the resonance of a nucleus other than that being investigated, e.g., the deuterium

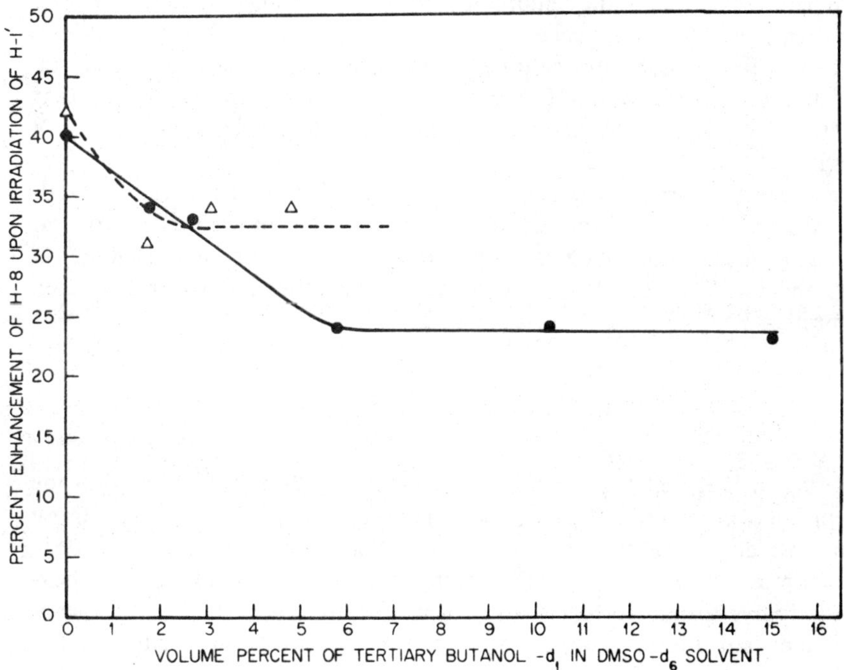

Fig. 5-1   Dependence of the NOE enhancement $f_8(1')$ on the concentration of t-butanol-d_1 in: ▲ 2',3'-isopropylidene–3,5'-cycloguanosine and ● 3,5'-cycloguanosine. The rf field was 2.78 mG. [Reprinted by permission from R. E. Schirmer, J. H. Noggle, J. P. Davis, and P. A. Hart, *J. Amer. Chem. Soc.* **92**, 3266 (1970).]

resonance of the solvent. If such spectrometers are not available, the lock sample can be placed in a sealed capillary tube within the sample tube or in the thin annulus formed by precision coaxial sample tubes. If an internal lock sample is used, its concentration should be kept as low as is consistent with a steady field lock. The lock sample resonance must be a sharp singlet with a resonance well removed from the spectral area under investigation. Common lock samples are TMS, chloroform, benzene, hexamethyldisilizane, hexa-methyldisiloxane, and t-butyl alcohol-d_1. If the lock sample is internal, it will reduce the NOE by providing an additional relaxation path for the protons whose NOE is being studied; the effect of lock sample concentration on NOE enhancements is shown in Fig. 5-1 (2). In addition, since the lock sample is being irradiated by the locking rf field, there is a distinct but unconfirmed possibility of an intermolecular NOE if internal lock is used. Lock samples which may associate with the compound being investigated should never be used internally. External locking removes all of these restrictions if the lock sample is physically removed from the analytical sample; coaxial or capillary lock removes all except the requirement that the resonance frequency be well removed from those frequencies being investigated. Internal deuterium lock seems to offer the best combination of convenient sample preparation and lack of interference with the NOE experiments. We have seen (Table 3-8) that H{D} NOE is insignificant and that D–H dipole–dipole interactions are small.

In order to permit facile reproduction of experimental results, all NOE investigations should report: details of the degassing procedure, purity and source of deuterated solvents, concentration of internal lock sample, if any, and the presence of suspended matter and other known or suspected impurities.

B. Sidebands and Modulation

In the ensuing sections we shall assume that the reader is familiar with the operation of commercial NMR spectrometers and shall emphasize only those points which are unique or especially important for NOE studies. Experimental measurement of the NOE requires the use of field-locked, frequency-sweep double resonance. Since this experimental technique requires the use of sidebands (sb), a little background on the subject seems appropriate.

For the purposes of NMR, sidebands can be produced by several methods; these include: frequency modulation of the rf, amplitude modulation of the rf, phase modulation of the rf, and magnetic field modulation. While the latter technique is by far the most common, amplitude modulation is similar and more easily analyzed. An rf oscillation produces a sinusoidal voltage of frequency† ω_r and amplitude A

$$V = A \sin\omega_r t \tag{5.1}$$

If the amplitude of the rf voltage is time dependent with a frequency ω_m (the modulation frequency)

$$A = a + b \sin\omega_m t$$

then

$$V = (a + b \sin\omega_m t)(\sin\omega_r t)$$

By the usual trigonometric relationships we get

$$V = a \sin\omega_r t + b \cos(\omega_r + \omega_m) t - b \cos(\omega_r - \omega_m) t \tag{5.2}$$

Thus, the modulated rf contains three frequencies, ω_r, $\omega_r + \omega_m$, and $\omega_r - \omega_m$, the latter two being called the sidebands. An NMR signal can be observed if *any* of the frequencies achieves the resonance condition, $\omega = \gamma H_0$. Typically, ω_m is an audio frequency (af) with $f_m \sim 10^3$ Hz and ω_r is a radio frequency (rf) with $f_r \sim 10^7$ Hz. Under practical conditions other frequencies will appear, e.g., second sidebands at $\omega_r \pm 2\omega_m$; this is especially true when field modulation is used.

If an NMR signal is observed directly with unmodulated rf and magnetic fields, the slow passage absorption mode line shape is given by the usual formula‡

$$v = \frac{\gamma H_r T_2 M_0}{1 + (\omega_r - \gamma H_0)^2 T_2^2 + S} \tag{5.3}$$

where H_r is the amplitude (in gauss) of the rf magnetic field and M_0 is the equilibrium magnetization. The saturation factor is defined as

$$S = \gamma^2 H_r^2 T_1 T_2 \tag{5.4}$$

Henceforth we shall express the dc magnetic field in frequency units as ω_0, which is defined as

$$\omega_0 \equiv \gamma H_0 \tag{5.5}$$

---

† Here, and throughout, the letter ω denotes the "frequency" in radians per second; it is 2π times the usual frequency (denoted f) in cycles per second or hertz (Hz); $\omega = 2\pi f$. Frequency in hertz is also commonly denoted with the greek letter v.

‡ More details are given by Bovey (*1b*, p. 16ff); the notational differences are slight. Also see Reference 6. Nearly all of the basic references cover this topic.

If the magnetic field is modulated at an audio frequency ω_m of amplitude H_m, the rf detected NMR signal (from the "rf unit") will contain both dc and ac components. If the ac components (at frequency ω_m) are then phase detected in an audio-frequency phase detector, the lineshape observed (af reference in phase, rf reference out of phase) is (*3*)

$$g = \frac{2^{-\frac{1}{2}}\gamma H_r\, T_2\, M_0\, J_0(\beta)\, J_1(\beta)}{1 + [\omega_0 - (\omega_r \pm \omega_m)]^2\, T_2^2 + J_1^2(\beta)\, S} \tag{5.6}$$

where $J_k(\beta)$ is the Bessel function of order k and $\beta = \gamma H_m/\omega_m$ is the modulation index. For the applications with which we shall be concerned, $\beta \ll 1$.

Compared to the direct absorption mode signal [Eq. (5.3)] the first sideband is reduced in amplitude by $J_0(\beta)$ (a negligible effect when $\beta \ll 1$); another factor of $2^{-\frac{1}{2}}$ reduces the first sideband signal for reasons explained by Anderson (*3*). Otherwise, the sideband NMR signal [Eq. (5.6)] is similar to the direct NMR absorption [Eq. (5.3)] with the amplitude of the rf field (H_r) being replaced by an *effective* rf field amplitude

$$H_{\text{eff}} = J_1(\beta)\, H_r \tag{5.7a}$$

For $\beta \ll 1$, $J_1(\beta) \sim \beta/2$, so the effective amplitude of the rf field in sideband absorption NMR is

$$H_{\text{eff}} \simeq \gamma H_r\, H_m/2\omega_m \tag{5.7b}$$

This H_{eff} will be called variously H_1, H_2, or H_L depending on whether the *observing*, *saturating*, or *locking* "rf fields," respectively, are being discussed.

For example, for the observing sideband (or loosely, the observing rf) we define

$$H_1 \simeq \gamma H_r\, H_m/2\omega_m$$

and obtain the lineshape

$$g_1 = \frac{2^{-\frac{1}{2}}\gamma H_1\, M_0\, T_2}{1 + (\omega_0 - \omega_1)^2\, T_2^2 + S_1} \tag{5.8}$$

where $\omega_1 = \omega_r \pm \omega_m$ depending on whether the "upper" or "lower" sideband is observed and the saturation factor [Eq. (5.4)] is redefined as

$$S_1 = \gamma^2\, H_1^2\, T_1\, T_2$$

(For the saturating field, we shall have $S_2 = \gamma^2\, H_2^2\, T_1\, T_2$.)

Equation (5.8) is, thus, nearly identical to Eq. (5.3), and the distinction between direct absorption with an rf field (ω_1, H_1)† and sideband spectroscopy with $\omega_1 = \omega_r \pm \omega_m$ and $H_1 = \gamma H_r H_m / 2\omega_m$ is slight; for most purposes they can be considered to be identical.

Equation (5.6) [or (5.8)], of course, predicts that two resonances will be observed; in field sweep NMR the resonant magnetic fields are $(\omega_1 = \omega_r + \omega_m)$

$$H_\ell = H_0 - \omega_m/\gamma \qquad \text{(lower sideband)}$$

and $(\omega_1 = \omega_r - \omega_m)$

$$H_u = H_0 + \omega_m/\gamma \qquad \text{(upper sideband)}$$

The *saturating* rf field produced by a strong (high H_m) modulation of the magnetic field will have an effective strength

$$H_2 \simeq \gamma H_r H_m / 2\omega_m$$

H_2 is usually varied by changing H_m rather than H_r, but its dependence on H_r and ω_m should be kept in mind.

There are numerous advantages to using audio frequencies as high as possible; in practice, an upper limit on the audio frequency is placed by the power required to create the necessary field amplitudes at high frequencies [Eq. (5.7)]. Aside from permitting field lock operation, sideband operation has the advantage of stabilizing the baseline and making signal amplitude measurements more reproducible.

A detailed review of the effects of modulation in magnetic resonance has been given by Haworth and Richards (4). This reference also discusses topics omitted here, such as the effect of altering the rf or audio phase and the appearance of "centerbands" in field modulated experiments.

C. Magnetic Field Lock

The lock signal is usually obtained from a sideband produced with a fixed frequency audio oscillator. The magnetic field is adjusted until one of the lock-sample sidebands (commonly the upper sideband) is at resonance. The audio phase detector for the lock frequency is adjusted so the reference voltage is out of phase with the audio signal from the rf unit. (The rf phase detector is in phase for all sidebands; the observing frequency af phase detector is in phase

---

† In most references, H_1 means the amplitude of the rf; here we reserve H_1 for the amplitude of the observing sideband because we will be concerned almost exclusively with sideband NMR.

[cf. Haworth and Richards (*4*), especially Fig. 3, p. 19].) Under these conditions the lock sample resonance will appear as a *dispersion mode*; i.e., the dc from the audio phase detector will be zero at the center of the resonance and will be positive at higher fields and negative at lower fields (or vice versa). If the field sweep is stopped when the lock signal is at resonance, the positive or negative voltage which result if the magnetic field drifts can be used in a feedback loop to correct the drift and stabilize the field. This feedback can be used to (a) change the rf frequency to follow the field drift and maintain the resonance condition; (b) cause the flux stabilizer to produce a change in the magnetic field equal and opposite to the drift, (c) cause the magnet power supply to alter the magnet current and return the field to resonance, or (d) any combination of the preceding.

It is interesting to note that the audio frequency voltage is a *maximum* at the center of the resonance—the dc from the audio phase detector is zero only because the audio reference is out of phase at that point. This audio-frequency voltage can be observed in various ways to monitor the field lock. A common method is to simply display the ac on an oscilloscope, perhaps as a Lissajous figure with the reference voltage. Alternatively, the ac can be detected with a diode detector producing a dc whose amplitude will be a maximum at resonance; this dc can be observed with a simple ammeter. The latter method has the advantage that dc ammeters are cheaper than oscilloscopes.

Aside from informing the operator when his field is locked, the diode detected dc can be used to optimize field homogeneity. A single field homogeneity parameter (usually the y axis gradient) can be optimized by adjustment until the lock signal is a maximum. Also the diode detected dc can be used in a feedback loop for automatic y gradient control.†

The amplitude (H_L) of the locking rf field should be kept at a minimum consistent with a stable field lock. The audio oscillator must be very stable since nothing compensates for its drift; it is possible that lock frequency drift is the limiting factor in locked field stability.

D. Homonuclear Double Resonance

A block diagram showing the essential features of a homonuclear NMDR spectrometer is given in Fig. 5-2. Three audio frequencies modulate the magnetic field:

(a) the observing field; audio frequency $f_1 = \omega_m(1)/2\pi$, effective amplitude H_1;
(b) the saturating field; audio frequency $f_2 = \omega_m(2)/2\pi$, effective amplitude H_2;
(c) the locking field; audio frequency $f_L = \omega_m(L)/2\pi$, effective amplitude H_L.

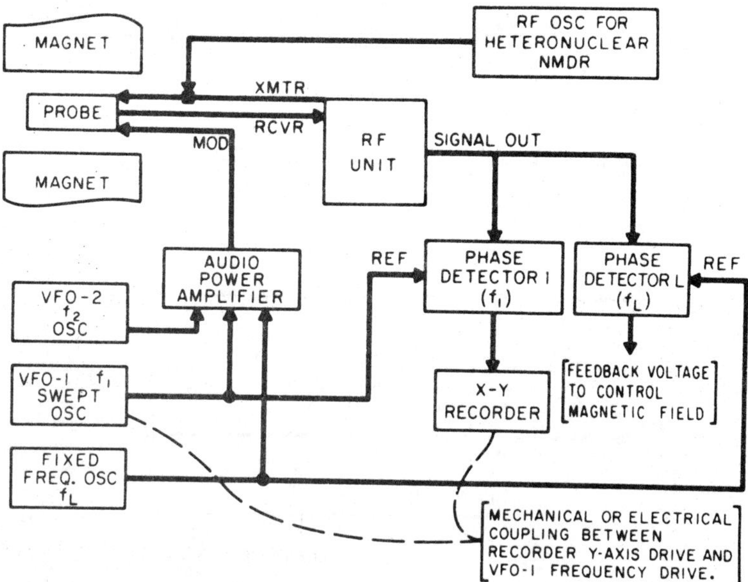

Fig. 5-2 A block diagram showing the essential features of a nuclear magnetic double resonance (NMDR) spectrometer. The following abbreviations are used: VFO, variable frequency oscillator; XMTR, transmitter; RCVR, receiver; REF, reference voltage to phase detectors; MOD, connection to the field modulation coils.

Additional variable frequency oscillators (VFO's) can be used for multiple resonance NMR. Phase detectors are provided at f_L for the field stabilization feedback and f_1 to observe the sideband NMR spectrum; f_2 is not detected.

The magnetic field can be locked on either the upper or lower sideband of the lock signal. (If one is initially observing the lock signal with the centerband rf, one sweeps upfield to the "upper" sideband, and vice versa.) The relationship between these sidebands and frequencies is shown in Fig. 5-3 using the familiar acetaldehyde and ethyl bromide spectra. If the upper sideband is used in conjunction with a "high field"‡ lock sample such as TMS (Fig. 5-3a), f_1 and f_2 will be higher than f_L and CHO will be at higher frequency than CH_3; i.e., "high field" resonances appear at the lower audio frequencies. On the lower sideband (Fig. 5-3b) the opposite prevails. This is illustrated for $CHO\{CH_3\}$ on acetaldehyde. On the Varian HA-100 spectrometer, typically $f_L = 2500$ Hz, and, routinely, frequency sweep spectra are observed on the upper sideband with f_1 swept from 2500–3500 Hz. This sweep range permits

---

† Such a system is used on many commercial spectrometers.

‡ "High field" is used in the conventional sense—the resonance at higher field in a field sweep experiment. TMS has the highest field proton spectrum commonly observed.

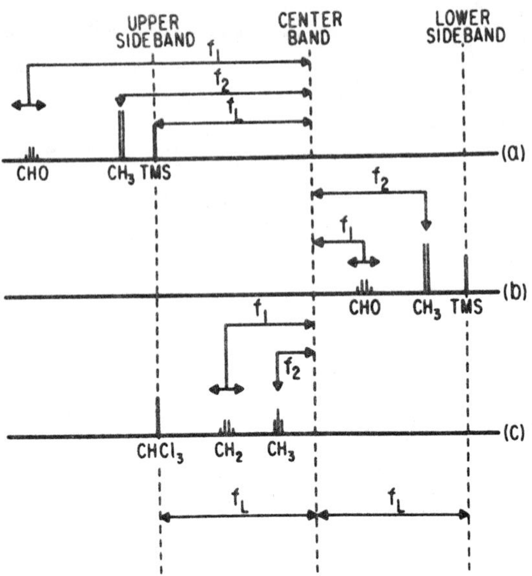

Fig. 5-3 Schematic presentation of frequency sweep, field locked, NMDR spectra for various conditions; (a) and (b), acetaldehyde with upper and lower sideband TMS lock respectively; (c), ethyl bromide with an upper sideband chloroform lock.

the recording of frequency sweep NMR spectra of protons over 10 ppm at 100 MHz (0–10τ if TMS lock is used).

It is often convenient or necessary to use a "low field" lock sample such as benzene or chloroform. In this case (cf. Fig. 5-3c), on upper sideband operation, the observing and saturating sidebands are lower in frequency than the lock sideband. If possible, it is a good idea in such cases to increase the frequency of the locking sideband so that the spectrum falls into the same frequency range as was the case with TMS lock.

The amplitudes of H_1 and H_2 can be adjusted more or less independently by adjusting their respective modulation amplitudes. There will, however, be some interaction between the sidebands; Bender (5) has shown that the effective amplitude of, for example, the first sideband at ω_1 is altered by another sideband at ω_2 by

$$H_1 = H_{\text{eff}} = J_1(\beta_1)\, H_r\, J_0(\beta_2)$$

[compare Eq. (5.7)] while the amplitude of an unsaturated resonance observed with the first sideband at ω_1 [Eq. (5.6)] is reduced by a factor of $J_0^2(\beta_2)$. Since $J_0 \to 1$ when $\beta \to 0$, these multiple sideband effects can be kept small by using a large rf amplitude (H_r) and small modulation indices (β). This may not always be practical so one must assume that gross changes in H_2 or ω_2 will alter the intensity of the observed signals. One important precaution is that equilibrium intensities should be measured by moving the saturating field (ω_2, H_2) off-resonance rather than turning off the H_2 modulation.

1. The Observing Field

The observing field H_1 should be kept small to avoid saturation. However, the effect of saturation on peak areas is less than on their amplitudes (6) so higher H_1's can be used if necessary. The observing audio frequency f_1 should be swept sufficiently slowly so that transient effects are avoided (e.g., "ringing" or asymmetric lineshapes). Excessive filtering should not be used. These conditions—nonsaturating slow passage with minimal filtering—are certainly safe in that systematic errors will be avoided. On the other hand, these conditions give very poor signal-to-noise and are probably never rigorously met in practical NMR [cf. Ernst (7)]. It is probable that many of the slow passage requirements can be violated without serious error since all that is required is that the area A of the resonance be such that $A \propto \langle I_z \rangle$. A systematic study of these effects would certainly be welcome. Ernst (7) gives a thorough treatment

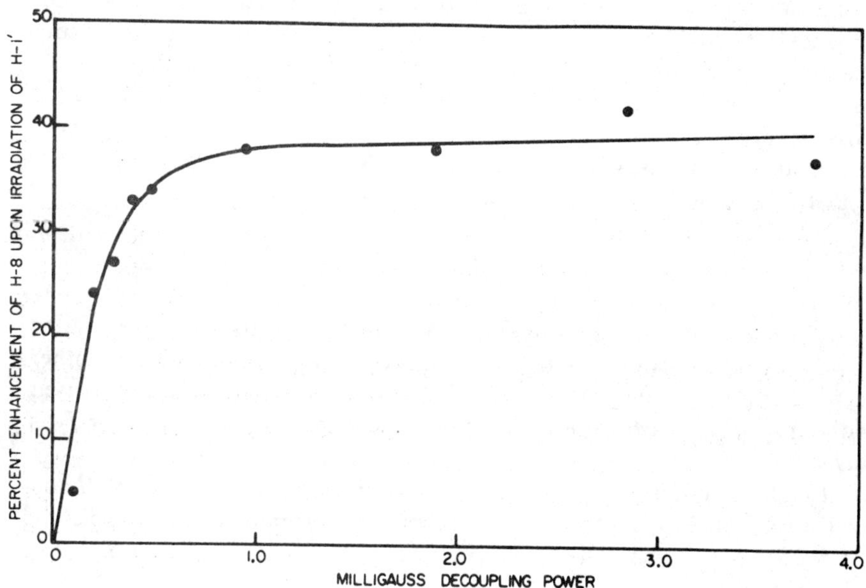

Fig. 5-4 Effect of the H_2 power level on the NOE enhancement $f_8(1')$ of 2',3'-isopropylidene-3,5'-cycloguanosine. The sample was a 0.25 M solution in DMSO-d_6. The lock sample was separated in a coaxial tube assembly. Reproduced with permission from Schirmer *et al.* (2).

of non-slow-passage solutions to the Bloch equations but gives too little attention to what errors are introduced in measured areas—a topic of prime concern with respect to the NOE.

The frequency f_1 (VFO-1) is commonly swept mechanically with a synchronous motor; electrical sweeps are possible.

2. The Saturating Field

The prime concern here is to cause M_z to be zero. Since f_2 is stationary, there is no approximation in the steady-state solution to the Bloch equations ($1b$) which gives

$$M_z = M_0 \frac{1 + T_2^2(\omega_0 - \omega_2)^2}{1 + (\omega_0 - \omega_2)^2 T_2^2 + S_2} \tag{5.9}$$

with

$$\omega_2 = \omega_r \pm 2\pi f_2 \quad \text{and} \quad S_2 = \gamma^2 H_2^2 T_1 T_2$$

When $\omega_0 = \omega_2$ and $S_2 \gg 1$, $M_z = 0$ and the maximum NOE is achieved; this is shown in Fig. 5-4 (2). If H_2 is made too large, the plateau value of the NOE enhancement could change for reasons which will be explained below. From the definition of S_2 it can be seen that the value of H_2 required to make $M_z = 0$ will vary with T_1 and T_2. It is desirable to minimize to any extent possible the required H_2 by maximizing T_1 and T_2, e.g., by using well-degassed, nonviscous, "clean" samples (cf. Section A, p. 96).

Being "on-resonance" simply requires that

$$(\omega_0 - \omega_2)^2 T_2^2 \ll 1 + S_2$$

Thus, the larger S_2, the less will be the effect of small drifts in f_2 or H_0 on the NOE enhancements. This effect is illustrated in Fig. 5-5 (2). On the other hand, there are various reasons why H_2 (and hence S_2) cannot be increased without limit so that the use of the most stable oscillator available for f_2 is advised.

At high power levels we must also be concerned about the effect of H_2 on nearby resonances. If another NMR absorption is off-resonance and sufficiently removed from the saturated resonance so that (for it) $(\omega_0 - \omega_2)^2 T_2^2 \gg 1$, the value of its M_z is

$$M_z(\text{off-resonance}) = M_0[1/(1 + x)] \tag{5.10}$$

where

$$x = S_2 /(\omega_0 - \omega_2)^2 T_2^2$$

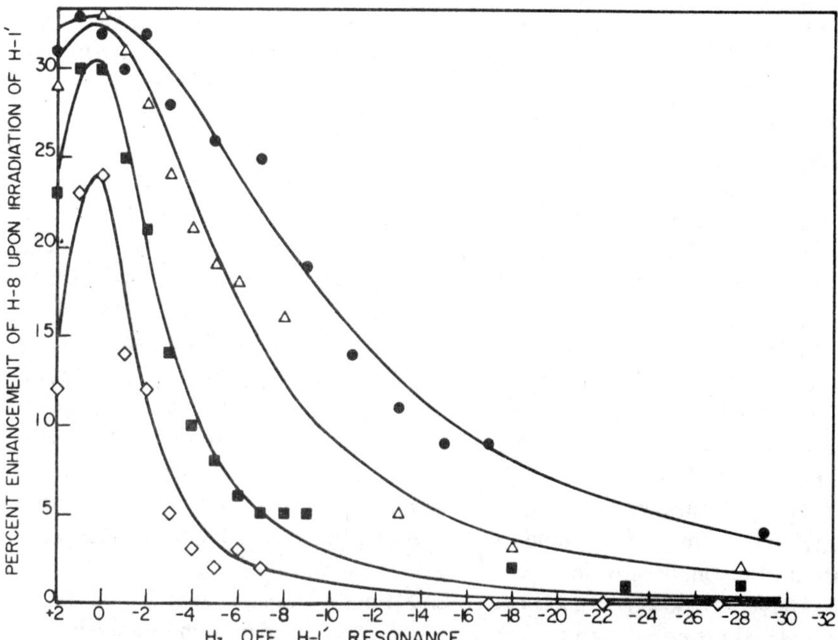

Fig. 5-5 Dependence of the observed NOE enhancement $f_8(1')$ of 2′,3′-isopropylidene-3,5′-cycloguanosine (0.25 M in DMSO-d_6 with 1.8% t-butyl alcohol-d_1, v/v) on the positioning of the irradiating field from the center of the 1′ resonance. The lines drawn are best fit Lorenzians; from Eq. (5.9) it can be seen that the deviation from maximum enhancement is approximately Lorenzian for small deviations of the irradiating sideband frequency from the center of the resonance. The effective field strengths are: ● $H_2 = 2.78$ milligauss; △ $H_2 = 1.86$ mG; ■ $H_2 = 0.94$ mG; ◇ $H = 0.5$ mG. [Reproduced by permission from R. E. Schirmer, J. H. Noggle, J. P. Davis, and P. A. Hart, *J. Amer. Chem. Soc.* **92**, 3266 (1970).]

For $M_z = M_0$ it is required that $x \ll 1$; i.e., the chemical shift δ in hertz between the saturated resonance and its nearest neighbor must obey

$$|\delta| \gg (\gamma/2\pi) H_2 (T_1/T_2)^{1/2} \tag{5.11}$$

Thus, very large values of H_2 may tend to saturate nearby resonances; this is one of the reasons that H_2 cannot be increased indefinitely. The effect is most serious if it occurs on the peak being observed, but even so the unintentional saturation of a neighboring peak can cause unanticipated Overhauser effects and introduce error into the interpretation of the NOE. If the condition in Eq. (5.11) is violated, a change could occur in the NOE enhancement plateau (Fig. 5-4) of the observed spin.

Another limitation on the size of H_2 is the possibility of receiver overload. The phase detector for f_1 is preceded by a wideband audio amplifier (necessarily wideband since f_1 is swept) which amplifies signals at f_2 as well as f_1. If the amplitude of the signal at f_2 is large, it may saturate the audio amplifier and cause anomalous effects on the signal at f_1, which of course gives the observed resonance. The result is an apparent decrease in the intensity of the signal being observed by f_1. This effect is particularly noticeable if the resonance being saturated is very strong. An analogous effect which occurs in cheap radio receivers is called *front end overload*. Kaiser (8) eliminated this effect by using a tunable notch filter to eliminate f_2 and pass f_1 prior to the phase detector f_1. This method will be effective only if f_2 and f_1 differ substantially (several hundred hertz or more) and may cause phase shifts in the resonances observed with f_1. Equally effective is to keep the audio and rf gain low before the detector stage; after detection f_2 is eliminated and the dc which results from f_1 can be amplified to any extent to drive the recorder or integrator.

NOE structural studies require that *all* components of a spin resonance including all lines of *J*-coupled multiplets be saturated. Therefore, the saturation of resonances which are spread out by *J* coupling will require proportionately more power. For very large *J* splittings, it may be advantageous to use several saturating sidebands so that the power required in each sideband is not excessive. For example, if a proton resonance is split by a 50-Hz fluorine spin coupling so that transitions occur at the audio frequencies 1250 and 1300 Hz, it would be better to use two saturating sidebands at these frequencies rather than a single, very strong, saturating sideband at 1275 Hz.

It is worth repeating that H_2 should never be turned off to measure equilibrium intensities but simply moved off resonance.

3. Calibration of H_2

If an NMR spectrum contains at least two chemically shifted peaks, there is an easy way to calibrate the strong rf amplitude H_2 (the units of H_2 are gauss or milligauss). If the strong field is centered on one of the resonances (perhaps by maximizing an NOE enhancement), it is an easy matter to measure the frequency ($f_1°$) at which the observing sideband is at the peak of the other resonance. This apparent chemical shift

$$\delta_{app} = f_2 - f_1° \tag{5.12}$$

differs from true chemical shift δ between the two peaks measured by the usual methods by an amount called the Bloch–Siegert shift (9). This shift is related to H_2 by

$$H_2 = (2\pi/\gamma)(\delta_{app}^2 - \delta^2)^{1/2} \tag{5.13}$$

Values of γ for a number of nuclei are given on Table 5-1. For protons, values of H_2 used for NOE studies will typically be several milligauss and $(\delta_{\text{app}} - \delta)$ will be several hertz. [Remember that H_2 is an "effective" field and depends on H_r and ω_2 as well as H_2; cf. Eq. (5.7).]

TABLE 5-1

GYROMAGNETIC RATIOS OF SOME COMMON NUCLEI[a]

Nucleus	γ(radians sec^{-1} G^{-1})	$\gamma/2\pi$(Hz G^{-1})	Spin I
^{1}H	26,752	4257.7	$\frac{1}{2}$
^{2}D	4,107	653.6	1
^{7}Li	$-1,040$	-1654.7	$\frac{3}{2}$
^{11}B	8,583	1366.0	$\frac{3}{2}$
^{13}C	6,726	1070.5	$\frac{1}{2}$
^{14}N	1,929	307.0	1
^{15}N	$-2,711$	-431.5	$\frac{1}{2}$
^{17}O	$-3,627$	-577.2	$\frac{5}{2}$
^{19}F	25,167	4005.5	$\frac{1}{2}$
^{23}Na	7,076	1126.2	$\frac{3}{2}$
^{27}Al	6,971	1109.4	$\frac{5}{2}$
^{29}Si	$-5,316$	-846.0	$\frac{1}{2}$
^{31}P	10,829	1723.5	$\frac{1}{2}$
^{33}S	2,052	326.6	$\frac{3}{2}$
^{35}Cl	2,621	417.2	$\frac{3}{2}$
^{37}Cl	2,181	347.2	$\frac{3}{2}$
^{39}K	1,248	198.7	$\frac{3}{2}$
^{51}V	7,033	1119.3	$\frac{7}{2}$
^{69}Ga	6,420	1021.8	$\frac{3}{2}$
^{71}Ga	8,158	1298.4	$\frac{3}{2}$
^{75}As	4,582	729.2	$\frac{3}{2}$
^{79}Br	6,702	1066.7	$\frac{3}{2}$
^{81}Br	7,224	1149.8	$\frac{3}{2}$
^{117}Sn	$-9,909$	$-1577.$	$\frac{1}{2}$
^{119}Sn	$-9,971$	$-1587.$	$\frac{1}{2}$
^{127}I	5,353	851.9	$\frac{5}{2}$
^{195}Pt	5,751	915.3	$\frac{1}{2}$
^{199}Hg	4,783	761.2	$\frac{1}{2}$
^{207}Pb	5,591	889.9	$\frac{1}{2}$

[a] Calculated from a table of nuclear properties published by Varian Associates, Palo Alto, California.

4. Effects of Audio Frequency Distortion

If three audio frequencies (f_1, f_2, and f_L) are provided to modulate the magnetic field, it does not necessarily follow that only these three frequencies will be present in the NMR spectrum. One of the results of multiple sideband theory (5) is that sidebands occur at many other frequencies, e.g., $m\omega_1 \pm n\omega_2$ m and n are small integers. These effects can also be caused by harmonic or intermodulation distortion. The stronger modulation at f_2 is particularly likely to produce such effects.

Harmonic distortion at f_2 produces $2f_2$, $3f_2$ etc., and therefore represents the same problem as higher-order sidebands. The experiment should be arranged so that f_1 and f_L are never at or near a harmonic of f_2. Likewise, if f_2 is a harmonic of f_1 or f_L, the phase detectors may respond to it. From Fig. 5-3 for proton–proton NMDR, we see that if one locks on the upper sideband of TMS, harmonics are unlikely to cause trouble if $f_L > 1000$ Hz. (the probable range of proton chemical shifts at 100 MHz.). If a "low field" lock (chloroform, benzene) is used harmonics are possible on upper sideband lock. These effects can be avoided however by increasing the lock frequency.

Intermodulation distortion produces frequencies such as $f_2 + f_1$, $f_2 - f_1$, $2f_2 - f_1$, etc. These could be particularly troublesome since so many combinations are possible. They can be discovered by the following procedure. Disconnect VFO-1 from the modulation coils and provide a fixed "f_1" with a manual oscillator; the amplitude and frequency should be typical of what VFO-1 has when it is used for modulation. VFO-1, which is still providing the reference voltage for phase detector 1, can then be swept through the entire "spectrum" without changing "f_1." No NMR spectrum will be observed, but when the reference frequency provided by VFO-1 passes through a frequency present in the rf unit output, e.g., "f_1," f_2, $2f_2$, etc., a beat pattern will be observed due to the fact that VFO-1 is not phase coherent with VFO-2 or the oscillator providing "f_1."† Likewise, if any intermodulation products are present they will show up as beat patterns. This "beat spectrum" can be compared with the usual NMR spectrum to see if any undesirable frequencies are located at critical spots.

Distortion effects can be eliminated by proper design of experiments and equipment.

5. Signal Integration

Since it is the areas of the NMR spectra which are of interest in NOE studies, the resonance must be integrated by some manner. Three methods are possible:

a. Electronic integration. Most spectrometers have an integrator for this

---

† These are exactly analogous to the beat patterns seen in ordinary homonuclear NMDR when f_1 is swept through f_2; the length of the beat is determined by the output filtering of the audio phase detector.

purpose. Integration of double resonance spectra may cause more difficulties than usual, especially if f_2 is near f_1. Baseline drift is, as always, a problem.

b. Mechanical integration. Spectra are recorded as absorption mode and integrated with a planimeter, by the cut-and-weigh method, or by counting squares on the graph paper.

c. Digital integration. If the spectrum is recorded on a digital signal averager or a computer, it can be integrated numerically. Some digital signal averagers† provide such a feature as a standard part of the instrument. This seems to be a worthwhile convenience for NOE studies.

E. Heteronuclear Double Resonance

Heteronuclear NMDR can be of two types: (a) massive decoupling of all nuclei of a given species, e.g., by modulating the saturating rf with white noise (*10*), and (b) selective decoupling. The latter type is of greatest interest for NOE studies. To obtain the stability required for selective decoupling the field-locked, frequency-sweep technique described above should be used. The only change in the experimental set-up (Fig. 5-2) is that VFO-2 is eliminated and the saturating rf field is supplied by a separate rf oscillator coupled into the transmitter (XMTR) coils in the probe. The source of the saturating rf should be a stable oscillator such as a frequency synthesizer with a power amplifier if necessary. About one watt of power is needed for T_1 and $T_2 \sim 1$ sec; more power may be needed for shorter relaxation times or if white noise modulation is used.

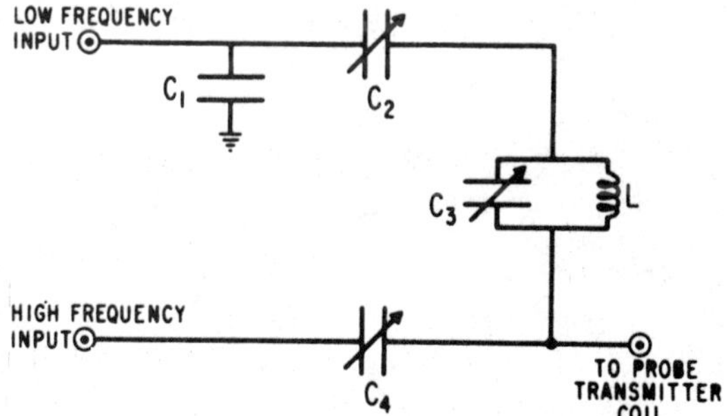

Fig. 5-6 A coupling circuit for double tuning the transmitter coils of an NMR probe. Reproduced with permission from Hopkins (*11*).

---

† For example, those manufactured by Nicolet Instruments, Inc. (formerly Fabri-tek), Madison, Wisconsin.

Usually, but not necessarily, the same transmitter coil (in the probe) is used for both the observing and saturating rf fields. A typical circuit which can be used to couple two frequencies into the same transmitter coil is shown in Fig. 5-6; this circuit, and details on its use and design are given by Hopkins (*11*).

If the spectrum of the nuclei being saturated was recorded previously by field sweep NMR, it is worth remembering that "high field" resonances will be at lower frequencies for NMDR.

REFERENCES

1a. E. D. Becker, "High Resolution NMR," Academic Press, New York, 1969.

1b. F. A. Bovey, "Nuclear Magnetic Resonance Spectroscopy," Academic Press, New York, 1969.

2. R. E. Schirmer, J. H. Noggle, J. P. Davis, and P. A. Hart, *J. Amer. Chem. Soc.* **92**, 3266 (1970).

3. W. A. Anderson, *Rev. Sci. Instrum.* **33**, 1160 (1962).

4. O. Haworth and R. E. Richards, *Progr. NMR Spectros.* **1**, 1 (1966).

5. P. Bender, private communication, 1970.

6. J. A. Pople, W. G. Schneider, and H. J. Bernstein, "High Resolution Nuclear Magnetic Resonance," McGraw-Hill, New York, 1959.

7. R. R. Ernst, *Advan. Magn. Resonance* **2**, 1 (1966).

8. R. Kaiser, *J. Chem. Phys.* **39**, 2435 (1963).

9. F. Bloch and A. Siegert, *Phys. Rev.* **57**, 522 (1940); W. A. Anderson, *in* "NMR and EPR Spectroscopy" (Varian Staff). Pergamon, New York, 1960.

10. R. R. Ernst, *J. Chem. Phys.* **45**, 3845 (1966).

11. R. C. Hopkins, Ph. D. Thesis, Harvard Univ. Cambridge, Massachusetts, 1965.

CHAPTER

6

TRANSIENT METHODS

In transient NMR experiments, the populations of the spin states are disturbed from their equilibrium or steady-state values and the decay of the system to a new steady-state is observed. The perturbation may be generated by (1) applying a strong rf field at the resonance frequency of one of the spins, (2) removing the strong field from a line which is already saturated, (3) by adiabatic rapid passage through the resonance, or (4) by application of a strong rf pulse at the resonance frequency. The behavior of a spin system under the various possible treatments is the subject of this chapter.

The general equation of motion for the expectation value $\langle I_{zi} \rangle$ of the z component of the ith nuclear spin operator has been given in Chapter 3 as [Eq. (3.1)]

$$d\langle \mathbf{I}_{zi} \rangle / dt = - R_i(\langle \mathbf{I}_{zi} \rangle - I_{0i}) - \sum_{j \neq i} \sigma_{ij}(\langle \mathbf{I}_{zj} \rangle - I_{0j}) \qquad (6.1)$$

The intensities measured in a transient experiment will be directly proportional to the $\langle I_{zi} \rangle$ and thus may be substituted for the $\langle I_{zi} \rangle$ in Eq. (6.1) providing, of course, that *all* $\langle I_{zi} \rangle$ are replaced by the corresponding intensities. The connection between experimental intensities and the quantities of interest, R_i and σ_{ij}, are obtained by solving Eq. (6.1) under boundary conditions appropriate to the experiment being performed.

113

For an N-spin system there will be N R's and $N(N-1)$ σ's ($\sigma_{ii} = 0$ for all i), or a total of N^2 quantities to be determined from the experimental data. In practice, systems with $N = 2$ are the most important and these will be considered first.

A. Experiments on Two-Spin Systems

In a two-spin system Eq. (6.1) becomes

$$d\langle \mathbf{I}_{zA}\rangle/dt = -R_A(\langle \mathbf{I}_{zA}\rangle - I_{0A}) - \sigma_{AB}(\langle \mathbf{I}_{zB}\rangle - I_{0B}) \tag{6.2}$$

$$d\langle \mathbf{I}_{zB}\rangle/dt = -R_B(\langle \mathbf{I}_{zB}\rangle - I_{0B}) - \sigma_{BA}(\langle \mathbf{I}_{zA}\rangle - I_{0A}) \tag{6.3}$$

The general solutions of (6.2) and (6.3) may be written

$$\langle \mathbf{I}_{zA}\rangle = I_{0A} + C_1 e^{-\lambda_1 t} + C_2 e^{-\lambda_2 t} \tag{6.4}$$

$$\langle \mathbf{I}_{zB}\rangle = I_{0B} + \frac{\lambda_1 - R_A}{\sigma_{BA}} C_1 e^{-\lambda_1 t} + \frac{\lambda_2 - R_A}{\sigma_{BA}} C_2 e^{-\lambda_2 t} \tag{6.5}$$

where λ_1 and λ_2 are roots of the characteristic equation.

$$\lambda_{1,2} = \tfrac{1}{2}\{(R_A + R_B) \pm [(R_A - R_B)^2 + 4\sigma_{AB}\sigma_{BA}]^{\frac{1}{2}}\} \tag{6.6}$$

Explicit expressions will now be presented for the constants in Eqs. (6.4) and (6.5) for several of the more common experiments on two-spin systems.

Experiment 1. *The Time Dependence of Signal A upon Instantaneous Saturation of Signal B*

At $t = 0$, a strong rf field is suddenly applied at the frequency of the B resonance. The field must be strong enough to saturate B essentially instantaneously. The strong rf field remains on throughout the experiment, and the A signal is monitored from time zero until a steady state is reached. The initial conditions required to evaluate the constants in Eqs. (6.4) and (6.5) are thus $\langle \mathbf{I}_{zA}(0)\rangle = I_{0A}$ and $\langle \mathbf{I}_{zB}(t)\rangle = \langle \mathbf{I}_{zB}(0)\rangle = 0$. The result is

$$\langle \mathbf{I}_{zA}\rangle = I_{0A} + (\sigma_{AB}/R_A) I_{0B}(1 - e^{-R_A t}) \tag{6.7}$$

The steady-state value of $\langle \mathbf{I}_{zA}\rangle$ is $I_{0A} + \sigma_{AB}I_{0B}/R_A$, in agreement with the discussion in the preceding chapters. The ratio of σ_{AB}/R_A can be obtained from $\langle \mathbf{I}_{zA}(\infty) \rangle$ and then R_A can be obtained from a plot of $\log[\langle \mathbf{I}_{zA}(t) - \langle \mathbf{I}_{zA}(\infty) \rangle]$ versus t, which will be a straight line with slope $-2.303R_A$. Repeating this procedure, but observing B while saturating A, yields R_B and σ_{BA} in a similar manner. All four dynamic constants characterizing the spin system are then known.

A practical problem arises in performing this experiment in that the saturation of the second line requires a finite time during which undesired transient effects will occur in the intensity of the observed line. These transients will decay to zero with a time constant of approximately T_2^*, the inverse of the experimental linewidth [see Eq. (1.4)]. This restricts the experiment to data points at times later than several T_2^*'s which in turn requires that the experiment be done on spins for which $R_i^{-1} \gg T_2^*$. Experiment 2 avoids this problem entirely.

Experiment 2.   *The Recovery of A from Saturation while B Remains Saturated*

Strong rf fields are applied at both the A and B frequencies for sufficiently long to saturate them both. The rf field at A is then removed and the recovery of A is followed by sweeping a weak observing field through it repetitively. The boundary conditions are thus $\langle \mathbf{I}_{zB}(t) \rangle = 0$ and $\langle \mathbf{I}_{zA}(0) \rangle = 0$. The solution of (6.2) and (6.3) is then

$$\langle \mathbf{I}_{zA}(t) \rangle = [I_{0A} + (\sigma_{AB}/R_A)\, I_{0B}](1 - e^{-R_A t}) \tag{6.8}$$

As with Experiment 1, σ_{AB}/R_A is obtained from I_{0B}, I_{0A}, and $\langle \mathbf{I}_{zA}(\infty) \rangle$, and R_A from a plot of $\log[\langle \mathbf{I}_{zA}(t) \rangle - \langle \mathbf{I}_{zA}(\infty) \rangle]$. Again, both R_A and σ_{AB} are obtained, and repeating the experiment with the roles of the spins interchanged results in determination of all four constants characterizing the dynamics of the spin system. This method has the advantages that the measured quantities vary over a wider range of values $[0 \leqslant \langle \mathbf{I}_{zA}(t) \rangle \leqslant I_{0A} + \sigma_{AB} I_{0B}/R_A$ as opposed to $I_{0A} \leqslant \langle \mathbf{I}_{zA}(t) \rangle \leqslant I_{0A} + \sigma_{AB} I_{0B}/R_A$ in Experiment 1], and the problem of initial transients is avoided.

Experiment 3.   *The Time Dependence of A after Removing the Saturating Field from B*

This experiment is performed by applying the strong rf field at the B resonance until a steady state is reached. At $t = 0$, the rf field is removed and the resonance at A is monitored as the spin system returns to equilibrium. The initial conditions are $\langle \mathbf{I}_{zB}(0) \rangle = 0$ and $\langle \mathbf{I}_{zA}(0) \rangle = I_{0A} + \sigma_{AB} I_{0B}/R_A$. These conditions result in

$$C_1 = \left[\frac{\sigma_{BA} - \sigma_{AB}}{\lambda_2 - \lambda_1} + \frac{\sigma_{AB}}{R_A} \frac{\lambda_2}{\lambda_2 - \lambda_1} \right] I_{0B}$$

$$(6.9)$$

$$C_2 = - \left[\frac{\sigma_{BA} - \sigma_{AB}}{\lambda_2 - \lambda_1} + \frac{\sigma_{AB}}{R_A} \frac{\lambda_1}{\lambda_2 - \lambda_1} \right] I_{0B}$$

with the λ_1 and λ_2 given in Eq. (6.6). The experiment is then repeated with B being observed after the saturating field is removed from A. The decay curves obtained in these experiments are fit to Eqs. (6.4) and (6.10), and Eqs. (6.11) are used to obtain the final parameter values

$$\langle \mathbf{I}_{zB} \rangle = I_{0B} + C_1{}^* e^{-\lambda_1 t} + C_2{}^* e^{-\lambda_2 t} \tag{6.10}$$

$$R_A = \frac{C_1 C_2{}^* \lambda_1 - C_1{}^* C_2 \lambda_2}{C_1 C_2{}^* - C_1{}^* C_2}, \qquad R_B = \lambda_1 + \lambda_2 - R_A$$

$$(6.11)$$

$$\sigma_{AB} = \frac{C_1 + C_2}{I_{0B}} R_A \qquad\qquad \sigma_{BA} = \frac{C_1}{C_1{}^*}(\lambda_1 - R_A)$$

It is apparent that this experiment will in general be more difficult to interpret than Experiments 1 or 2 because it will require fitting a two-exponential equation to the decay curve. The problem of fitting double or multiple exponentials to an observed decay curve is discussed briefly in Appendix II. Special conditions existing in some applications ($I_{0A} = I_{0B}$, $R_A = R_B$, etc.) may make this experiment much more easily interpreted [cf. Anet and Bourn (*1*)].

Experiment 4. *The Time Dependence of A after Inverting the Magnetization at B*

At time $t = 0$, the magnetization of spins B is inverted by applying a selective 180° rf pulse or by an adiabatic fast passage. The initial conditions are then $\langle \mathbf{I}_{zA}(0) \rangle = +I_{0A}$ and $\langle \mathbf{I}_{zB}(0) \rangle = -I_{0B}$. With these boundary conditions the constants in Eqs. (6.4) and (6.5) become

$$C_1 = -C_2 = -2I_{0B}\sigma_{BA}/(\lambda_1 - \lambda_2) \tag{6.12}$$

with λ_1 and λ_2 again given by (6.6). Repeating the experiment by observing B after inverting A and fitting the decay to Eq. (6.10), the R's and σ's can be obtained from

$$R_A = \frac{\lambda_1 + \lambda_2}{2} + (\lambda_1 + \lambda_2)\left(1 - \frac{C_2 C_2^*}{I_{0A} I_{0B}}\right)^{\frac{1}{2}}, \qquad R_B = \lambda_1 + \lambda_2 - R_A$$

$$\sigma_{BA} = C_2(\lambda_1 - \lambda_2)/2I_{0B}, \qquad \sigma_{AB} = C_2^*(\lambda_1 - \lambda_2)/2I_{0A} \tag{6.13}$$

where C_2^* is defined by Eq. (6.10).

Experiment 5. *The Time Dependence of A after Inverting the Magnetization at A*

At $t = 0$, the magnetization of spin A is inverted by applying a selective $180°$ rf pulse or by an adiabatic fast passage. The initial conditions are then $\langle I_{zA}(0)\rangle = -I_{0A}$ and $\langle I_{zB}(0)\rangle = +I_{0B}$. Using these boundary conditions in Eqs. (6.4) and (6.5) we obtain

$$C_1 = 2I_{0A}\frac{\lambda_2 - R_A}{\lambda_1 - \lambda_2}, \qquad C_2 = -2I_{0A}\frac{\lambda_1 - R_A}{\lambda_1 - \lambda_2} \tag{6.14}$$

with λ_1 and λ_2 being given by Eq. (6.6). This experiment, including several special cases, has already been discussed in Chapter 1, Section G.

Experiment 6. *The Return of A to Equilibrium after Saturating A*

After placing a strong rf field at the A resonance for a sufficient time to allow the system to reach a steady state with A completely saturated, the rf field is removed at $t = 0$ and the return of A and B to equilibrium is monitored. The initial conditions are $\langle I_{zA}(0)\rangle = 0$ and $\langle I_{zB}(0)\rangle = I_{0B} + \sigma_{BA} I_{0A}/R_B$. The constants then become

$$C_1 = I_{0A}\left(\frac{\lambda_2 - R_A}{\lambda_1 - \lambda_2}\right) + I_{0B}\frac{\sigma_{BA}}{R_B}\left(\frac{\sigma_{BA} - R_B}{\lambda_1 - \lambda_2}\right)$$

$$C_2 = -I_{0A}\left(\frac{\lambda_1 - R_A}{\lambda_1 - \lambda_2}\right) - I_{0B}\frac{\sigma_{BA}}{R_B}\left(\frac{\sigma_{BA} - R_B}{\lambda_1 - \lambda_2}\right) \tag{6.15}$$

and the λ's are given by Eq. (6.6).

Experiment 7. *The Return of A to Equilibrium after Saturating both A and B*

This experiment consists of saturating both the A and B resonances by placing a strong rf field at each (or a single strong field in the vicinity of both if A and B are closely spaced resonances) and suddenly removing the strong field(s) at $t = 0$. The return of both A and B to equilibrium may be monitored simultaneously. The initial conditions in this case are $\langle I_{zA}(0) \rangle = \langle I_{zB}(0) \rangle = 0$, and the resulting constants are

$$C_1 = I_{0A}\left(\frac{\lambda_2 - R_A}{\lambda_1 - \lambda_2}\right) - I_{0B}\left(\frac{\sigma_{BA}}{\lambda_1 - \lambda_2}\right)$$

$$C_2 = -I_{0A}\left(\frac{\lambda_1 - R_A}{\lambda_1 - \lambda_2}\right) - I_{0B}\left(\frac{\sigma_{BA}}{\lambda_1 - \lambda_2}\right)$$

$$(6.16)$$

with λ_1 and λ_2 given by Eq. (6.6).

B. Experiments on Multiple-Spin Systems

The solution of Eqs. (6.1) will, in general, be a sum of N exponentials for any one of the N types of spin in a multiple-spin system. In accordance with this, the time dependence observed in experiments on two-spin systems would generally consist of the sum of two exponentials. In Experiments 1 and 2 of the preceding section, however, only a single exponential resulted. The reason for this was that all spins except the observed one were saturated throughout the course of the experiment, so that their presence was reflected only in the terms R_i and $\sigma_{ij}I_{0j}$ in the equation of motion of the observed spin. The additional exponentials that would result from the time dependence of the other $\langle I_{zi}(t) \rangle$ were suppressed by the choice of experimental conditions. The behavior of any multispin system can be reduced to a single exponential in this fashion.

As was first pointed out by Forsén and Hoffman (2), the values of the R's and σ's for a system of more than two spins are most easily extracted by a combination of transient and steady-state experiments. Both the transient and supplementary steady-state experiments will be discussed in this section. An alternative procedure using a nonspecific 180° rf pulse and Fourier transform of the resulting decay curve, but not employing the simplifications characteristic of the multiple resonance experiments, will be deferred to the following section.

Experiment 8   *The Time Dependence of A When All Other Resonances Are Saturated at $t = 0$*

In this experiment, the frequencies and strengths of the oscillators are set so that, when the power is turned on, all lines in the spectrum except A will be rapidly and completely saturated. At time $t = 0$, the power is turned on and the intensity of resonance A is observed as a function of time by repetitively sweeping a weak observing rf field through it. The appropriate boundary conditions to be used in solving Eq. (6.1) for this experiment are

$$\langle \mathbf{I}_{zj}(0)\rangle = \begin{cases} 0, & j \neq A \\ I_{0A}, & j = A \end{cases}$$

Substituting these conditions into (6.1) and solving

$$\langle \mathbf{I}_{zA}(t)\rangle = I_{0A} + R_A^{-1}\left(\sum_{j \neq A} \sigma_{Aj} I_{0j}\right)(1 - e^{-R_A t}) \tag{6.17}$$

As $t \to \infty$, a steady state is approached in which $\langle \mathbf{I}_{zA}\rangle = I_{0A} + R_A^{-1}\sum \sigma_{Aj} I_{0j}$. $\langle \mathbf{I}_{zA}(\infty)\rangle$ and I_{0A} are easily measured, which allows $R_A^{-1}\sum_{j \neq A} \sigma_{Aj} I_{0j}$ to be calculated. A plot of $\log\left[-\langle \mathbf{I}_{zA}(t)\rangle + I_{0A} + \sum \sigma_{Aj} I_{0j}/R_A\right]$ versus t then gives a straight line of slope $-2.303 R_A$, so that both R_A and $\sum \sigma_{Aj} I_{0j}$ are obtained from this experiment. In an ideal case where this experiment can be performed on all the spins, the N values of the R_A and N relations of the form

$$\sum_{j \neq A} \sigma_{Aj} I_{0j} = \text{constant}$$

are obtained. Thus $N(N-2)$ additional relations are required to completely determine the σ's. Experiment 10 may be used to supply the additional relationships.

Again, as with Experiment 1, undesired transients will occur for a short time after the rf power is turned on, so that data collected at very early times must be discarded.

Experiment 9. *The Recovery of A from Saturation while the Other Resonances Remain Saturated*

Strong rf fields are applied at the frequencies of all lines in the spectrum, including A, until all the lines are saturated. At $t = 0$, the rf field at the A resonance is turned off, and the recovery of A is monitored by repetitively sweeping the weak observing field through it. The boundary conditions are $\langle \mathbf{I}_{zk}\rangle = 0$ at all times for $k \neq A$, and $\langle \mathbf{I}_{zA}(0)\rangle = 0$. The solution of Eq. (6.1) under these conditions is

$$\langle \mathbf{I}_{zA}(t)\rangle = \left(I_{0A} + \sum_{j \neq A} \sigma_{Aj} I_{0j}\Big/R_A\right)(1 - e^{-R_A t}) \tag{6.18}$$

Again, the determination of I_{0A} and $\langle \mathbf{I}_{zA}(\infty)\rangle$ followed by a plot of $\log[\langle \mathbf{I}_{zA}(t)\rangle - I_{0A} - \sum \sigma_{Aj} I_{0j}]$ versus t will allow both R_A and $\sum \sigma_{Aj} I_{0j}$ to be determined. This procedure has the advantage that the problem of initial transients is avoided.

Experiment 10.   *A Steady-State Experiment: Measuring the Intensities of Lines A and B with All Other Resonances Saturated*

Strong rf fields are impressed at the frequencies of all lines in the spectrum except A and B, and the intensities of lines A and B are measured. The conditions used in solving Eq. (6.1) are $d\langle \mathbf{I}_{zk}\rangle/dt = 0$ for all k, and $\langle \mathbf{I}_{zk}(t)\rangle = 0$ for all $k \neq A, B$. This reduces Eqs. (6.1) to

$$- R_A(\langle \mathbf{I}_{zA}\rangle - I_{0A}) - \sigma_{AB}(\langle \mathbf{I}_{zB}\rangle - I_{0B}) + \sum_{j \neq A,B} \sigma_{Aj} I_{0j} = 0$$

$$- R_B(\langle \mathbf{I}_{zB}\rangle - I_{0B}) - \sigma_{BA}(\langle \mathbf{I}_{zA}\rangle - I_{0A}) + \sum_{j \neq A,B} \sigma_{Bj} I_{0j} = 0$$

which can be solved to give

$$\sigma_{AB} = -(\langle \mathbf{I}_{zB}\rangle - I_{0B})^{-1}\left[R_A(\langle \mathbf{I}_{zA}\rangle - I_{0A}) + \sum \sigma_{Aj} I_{0j}\right]$$

$$\sigma_{BA} = -(\langle \mathbf{I}_{zA}\rangle - I_{0A})^{-1}\left[R_B(\langle \mathbf{I}_{zB}\rangle - I_{0B}) + \sum \sigma_{Bj} I_{0j}\right]$$

$$(6.19)$$

In favorable cases, where the experiments can be done on all spins, values of R_A and R_B obtained independently from Experiments 8 or 9 can be combined with the results of Experiment 10 [using Eqs. (6.19)] to obtain values for all kinetic parameters of the spin system. In fact, if all conceivable experiments of each type could be performed, the parameters would be overdetermined. As mentioned earlier, there are N^2 parameters to be determined. Experiments 8 or 9 provide $2N$ relations among these constants and Experiments 10 provide $N(N-1)$ more, for a total of $N^2 + N$ independent relations. Thus the constants are all determined and are subject to N independent cross checks.

C. The Fourier Transform Method

The Fourier transform method is based upon the fact that the response of a nuclear spin system to a rf pulse contains the same information as a high-resolution spectrum taken just preceding that pulse. This follows from the proof by Lowe and Norberg (*3*) that the free induction decay and high resolution spectrum are just Fourier transforms of one another. This is a generalization of the well-known theorem that the frequency response function of a linear system and the unit impulse response form a Fourier transform pair (*4*).

The fact that, in practice, the Fourier transform technique offers significant advantages over the usual frequency or field-swept method was first pointed out by Ernst and Anderson (*5*). Several of the advantages they demonstrated are:

(1) The time required to obtain a spectrum is much less than that required by conventional sweep procedures, being on the order of $1/r$ seconds for a single sweep, where r is the resolution in hertz to be achieved in the experiment. The Fourier transform technique is thus more amenable to studies of time dependent phenomena such as chemical exchange or spin relaxation.

On the other hand, as a single free-induction decay can be obtained so rapidly by this technique, it is quite practical to repeat the experiment numerous times in a given period and add the signals in a time averaging computer to improve the signal-to-noise ratio, just as is done in high-resolution NMR. However, the time required to obtain a given signal-to-noise ratio by the Fourier transform technique will be less by a factor of $\Delta v/r$ than that required to attain the same ratio by time averaging the high-resolution spectrum. Δv is the width of the spectrum and would be on the order of a few hundred hertz for ^{1}H or a few thousand hertz for ^{13}C.

Ernst and Anderson (*5*) obtained values of $\Delta v/r$ on the order of 100 for protons at 60 MHz, and values on the order of 3000 should be attainable for ^{13}C at 15 MHz (note that $\Delta v \propto H_0$ so higher field strengths increase the ratio $\Delta v/r$). Several schemes using special sequences of rf pulses to obtain even greater time savings have been suggested (*6–8*).

(2) The second advantage of the Fourier transform method is that its sensitivity is greater than that of the swept experiment, providing that there is sufficient fine structure in the spectrum. In a given period of time, the sensitivity obtained using the transform method will be $(\Delta v/r)^{1/2}$ greater than that obtained by sweep procedures.

(3) Accurate frequency calibration of the spectrum is simplified and requires only accurate measurements of time in the experiment.

The procedure used to obtain a high-resolution spectrum of a sample using the Fourier transform technique is as follows. At time $t = 0$ a nonselective 90° rf pulse is applied to the sample. This requires an rf field strength satisfying $\gamma H_1/2\pi \gg \Delta v$, and a pulse duration on the order of $1/\gamma H_1$ (*9*). The effect of this pulse is to tip the magnetization $\gamma \hbar \langle \mathbf{I}_{zi}(0) \rangle$ into the xy plane where it is detected and its decay to zero recorded, generally by using a time-averaging computer. The Fourier transform of the signal is computed (see Appendix II) and yields the high-resolution spectrum.

The decay curve itself will contain components oscillating at frequencies corresponding to the chemical shifts and coupling constants present in the high-resolution spectrum of the sample. Figure 6-1b shows the complex decay curve obtained by Farrar (7) from the ^{13}C in CH$_3$Cl—a relatively simple system by most standards. The transformed spectrum is shown in Fig. 6-1a.

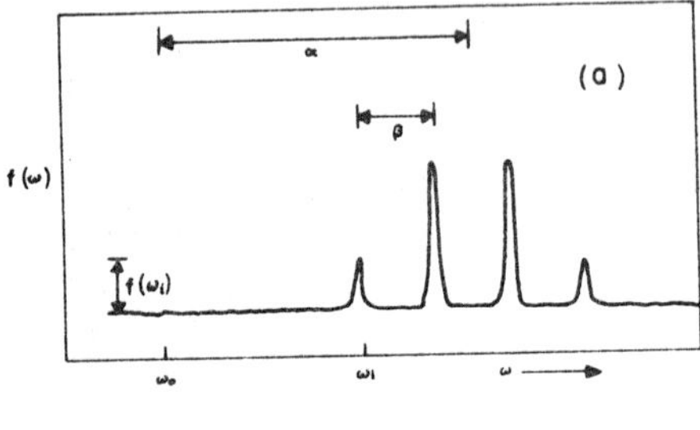

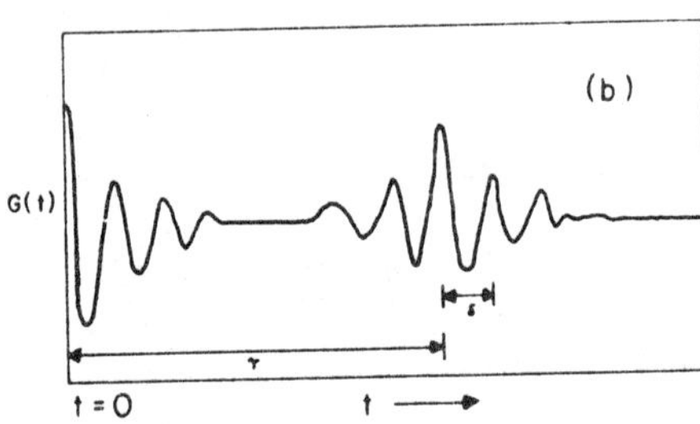

Fig. 6-1   (a), The normal ^{13}C high-resolution NMR spectrum of H$_3$CCl. α is the chemical shift of the quartet relative to the reference frequency ω_0. β is the ^{13}C–H spin coupling constant. (b), The NMR signal of the ^{13}C in H$_3$CCl observed following the rf pulse in a Fourier transform experiment. The ^{13}C–H coupling constant β is proportional to $1/\gamma$, and the chemical shift α is proportional to $1/\delta$. The spectra depicted in (a) and in (b) are interconvertible via the Fourier transform (7).

Unlike other pulsed NMR experiments, Fourier transform experiments do require a high-resolution magnet. The reason is that the ultimate resolution in the transformed spectrum will be determined by field inhomogeneity, just as with high-resolution spectra. It should be kept in mind, however, that the ultimate sensitivity will not be attained in the transformed spectrum unless the time averaging computer has at least 2 · (total sweep width)/(line width) channels.

Fourier transform methods, with the exception of the procedure discussed by Allerhand and Cochran (8), have the problem that correct adjustment of the phase to get a pure absorption mode spectrum is very difficult experimentally. It is generally easier to accept poor phase in the experiment and then correct it mathematically (5, 10) while analyzing the data. One procedure for correcting the phase (5) is to compute both the sine $\mathscr{S}(\omega)$ and cosine $\mathscr{C}(\omega)$ transform of the observed decay envelope. The pure absorption mode signal will be given by some linear combination of these two transforms,

$$I(\omega) = \mathscr{C}(\omega) \cos \phi + \mathscr{S}(\omega) \sin \phi$$

where ϕ is the error in the experimental phase angle. The correct value of ϕ can be determined by maximizing the ratio

$$R(\phi) = \sum_{i=1}^{N} (I_{\max} - I_i) \Big/ \sum_{i=1}^{N} (I_i - I_{\min})$$

where $I_{\max}$ is the greatest amplitude found in $I(\omega)$ and $I_{\min}$ is the least. The procedure is to compute values of $\mathscr{C}(\omega)$ and $\mathscr{S}(\omega)$ at N points using the observed decay curve. A value of ϕ is then selected and a set of values of $I(\omega)$ computed using that ϕ. Finally, $R(\phi)$ is computed. This is repeated systematically with other values of ϕ until $R(\phi)$ is maximized, in which case $I(\omega)$ is the pure absorption mode spectrum. Ernst has also discussed the use of Hilbert transforms to accomplish the phase adjustment (*10*).

In the case that we wish to study the spin–lattice relaxation of a spin system the Fourier transform method offers additional advantages because the relaxation of all lines in the spectrum can be studied simultaneously (*11*). The procedure employed in this case is to apply a nonselective 180° pulse at $t = 0$. The effect of the pulse is to set $\langle I_{zi}(0)\rangle = -I_{0i}$ for every line i of the spectrum, or, in terms of the high-resolution spectrum, to turn the spectrum upside down as is shown in Fig. 6-2. The magnetizations begin to decay toward their equilibrium values of $+I_{0i}$ due to the usual relaxation processes. At a time $t = \tau$ later a nonselective 90° pulse is applied to the sample and the decay of the NMR signal following the pulse is recorded. The Fourier transform is computed and yields the spectrum as it looked at $t = \tau$. The spectrum will,

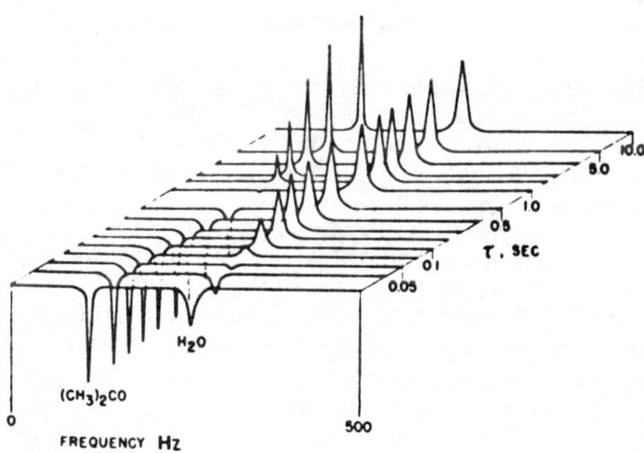

Fig. 6-2 The longitudinal proton relaxation surface for a 1.5×10^{-3} M solution of $FeCl_3$ in a 1:1 water:acetone mixture. Each spectrum results from the Fourier transform of the free induction decay following the second pulse of a 180°–90° pulse sequence. The time interval between pulses is indicated along the right-hand edge of the spectrum. The spectra are displaced isometrically on a logarithmic scale. Reproduced by permission from Vold *et al.* (*11*).

of course, be intermediate between the inverted spectrum generated by the pulse at $t = 0$ and the normal upright spectrum to which the system relaxes as $t \to \infty$ (see Fig. 6-2). The experiment is repeated for various values of τ, resulting in a complete characterization of the relaxation behavior of each line in the spectrum. The Eqs. (6.1) may be fitted to the decay curves and the σ's and R's extracted.

Fitting a system of equations to a series of multiexponential curves is a rather complicated and hazardous undertaking (see Appendix II). The motivation behind many of the Experiments 1–9 was, in fact, to avoid this problem. The interpretation of the data in those experiments was greatly simplified by forcing many of the experimental quantities to be zero and constant throughout the experiment by saturating their respective resonances. The curves obtained under these conditions contain only one or two exponentials and can be analyzed fairly easily by hand.

The same technique can be applied to a Fourier transform experiment to simplify its interpretation. Strong rf fields are applied at the resonance frequencies of all lines which are to be saturated and these remain on throughout the experiment, just as was done in the high-resolution experiments. The interpretation of the results would be the same as the analogous high resolution experiment. Freeman (*12*) has reported an experiment of this type and presented a brief discussion of the experimental technique including a method for stabilizing the field-to-frequency ratio.

REFERENCES

1. F. A. L. Anet and A. J. R. Bourn, *J. Amer. Chem. Soc.* **89**, 760 (1967).
2. S. Forsén and R. Hoffman, *J. Chem. Phys.* **40**, 1189 (1964).
3. I. J. Lowe and R. E. Norberg, *Phys. Rev.* **107**, 46 (1957).
4. V. V. Solodovnikov, "Introduction to the Statistical Dynamics of Automatic Control Systems." Dover, New York, 1960.
5. R. R. Ernst and W. A. Anderson, *Rev. Sci. Instrum.* **37**, 93 (1966).
6. E. D. Becker, J. A. Ferretti, and T. C. Farrar, *J. Amer. Chem. Soc.* **91**, 7784 (1969).
7. T. C. Farrar, *Anal. Chem.* **42**, No. 4, 110A (1970).
8. A. Allerhand and D. W. Cochran, *J. Amer. Chem. Soc.* **92**, 4482 (1970).
9. A. Abragam, "The Principles of Nuclear Magnetism," Chapter 2, Sect. E. Oxford Univ. Press, London and New York, 1961.
10. R. R. Ernst, *J. Magn. Resonance* **1**, 7 (1969).
11. R. L. Vold, J. S. Waugh, M. P. Klein, and D. E. Phelps, *J. Chem. Phys.* **48**, 3831 (1968).
12. R. Freeman, *J. Chem. Phys.* **53**, 457 (1970).

THE EFFECTS OF CHEMICAL EXCHANGE

The effects of chemical exchange (*1*) on a spin system may be somewhat arbitrarily divided into direct and indirect effects. The direct effect is the transfer of magnetization between the various sites. If, for example, the magnetization in one site were forced to zero by saturating the resonance, there would be a tendency for the magnetization in all other sites connected to it by exchange to be reduced due to the direct transfer of nuclei (and thus of magnetization) between sites. The indirect effect is the spin–lattice relaxation which may be induced by exchange. This arises because the interactions between spins become random functions of time in the presence of exchange. Both direct and indirect effects may, of course, be present simultaneously.

Most of the Overhauser studies on exchanging systems which appear in the literature are primarily concerned with the direct effects. The reason for this is that the direct effects provide a means for studying the exchange process itself in a range of slow exchange rates not accessible by other NMR techniques. However, even though they are not of primary interest, indirect effects must often be taken into account in interpreting the experimental results. Because of this, the indirect effects will be discussed first in this chapter and then we shall proceed to more general cases, including a review of selected literature.

A. Chemical Exchange Induced Spin–Lattice Relaxation

We found earlier that all relaxation pathways available to a spin contributed to the value of R for that spin. The greater the number of pathways, the greater R becomes and the smaller the magnitude of the Overhauser effect becomes. It is for this reason that relaxation induced by chemical exchange becomes important in interpreting the results of an Overhauser experiment.

Spin–lattice relaxation in spin-$\frac{1}{2}$ systems is the result of transitions between spin states induced by random fluctuations $H'(t)$ in the magnetic fields experienced by the nuclei. The effects of fluctuations in the x and y components of the fields must, however, be distinguished from the effects of fluctuations in the z component, i.e., in the component parallel to the static field H_0. The z component, $H_z'(t)$, may contribute to the decay of the transverse magnetization ($\langle I_{ix} \rangle$ and $\langle I_{iy} \rangle$) but will not cause changes in $\langle I_{iz} \rangle$ (2). $H_x'(t)$ and $H_y'(t)$, on the other hand, cause changes in $\langle I_{iz} \rangle$ as well as $\langle I_{ix} \rangle$ and $\langle I_{iy} \rangle$ and thus do contribute to spin–lattice relaxation. Because of this, we can determine whether or not it is possible for exchange modulation of a given interaction to provide a relaxation path for the spin simply by deciding whether or not the interaction has nonzero x and y components.

The Hamiltonian for a spin-$\frac{1}{2}$ system typically contains three groups of terms: Zeeman or chemical shift terms, scalar couplings, and dipole–dipole interactions. The chemical shift terms are of the form $\gamma H_0 I(1 - \sigma)$ and have no x or y components. They cannot, therefore, contribute to spin–lattice relaxation. It might be noted that the presence of anisotropy in the chemical shift could alter this conclusion as the anisotropy would result in x and y components in the Zeeman coupling (3). There are, however, no cases known to us where chemical shift anisotropy is a significant relaxation mechanism (see Section 2,I), and this unlikely case will not be considered here.

The scalar coupling, $J\mathbf{I}\cdot\mathbf{S} = J(\mathbf{I}_x\mathbf{S}_x + \mathbf{I}_y\mathbf{S}_y + \mathbf{I}_z\mathbf{S}_z)$, does contain x and y components and can contribute to spin–lattice relaxation when modulated either by rapid relaxation of one spin [$S = S(t)$] or by chemical exchange [$J = J(t)$]. The former case has already been discussed in Chapter 2, Section G. The effects of exchange modulation have been computed by Abragam†† with the result that

$$\rho_{IS} = \frac{8\pi^2 J^2}{3} \frac{\tau_{exch}}{1 + (\omega_I - \omega_S)^2 \tau_{exch}^2} S(S + 1) \tag{7.1}$$

$$\sigma_{IS} = -\frac{8\pi^2 J^2}{3} \frac{\tau_{exch}}{1 + (\omega_I - \omega_S)^2 \tau_{exch}^2} I(I + 1) \tag{7.2}$$

---

† See Abragam (3, p. 308).

where J is the coupling constant, S and I are the spins of nuclei S and I ($S = I = \frac{1}{2}$ in most cases of interest here), and τ_{exch} is the exchange correlation time. The correlation time is related to the first-order or pseudo-first-order rate constant for the exchange process by $1/k = \tau_{exch}$. Interchanging I and S in (7.1) and (7.2) yields the corresponding expressions for σ_{SI} and ρ_{SI}.

Equation (7.1) shows that scalar relaxation of I by S will tend to zero for very fast or very slow exchange, and will be maximum when $k^2 = (\omega_I - \omega_S)^2$. The maximum value is thus

$$\rho_{IS} = \frac{4\pi^2 S(S+1) J^2}{3|\omega_I - \omega_S|}, \qquad k = |\omega_I - \omega_S|$$

To get some feel for the magnitude of this quantity, suppose I and S are protons with a relative chemical shift of 100 Hz. If $J = 1$ Hz, then $\rho_{IS} = 0.016$ sec^{-1}. If $J = 10$ Hz, $\rho_{IS} = 1.6$ sec^{-1}. These should be compared with values of $R_I = \sum_j \rho_{Ij} = 0.1$ sec^{-1} typically found for protons in solution.

The maximum fractional enhancement of I due to saturation of S can also be computed from Eqs. (7.1) and (7.2).

$$f_I(S)_{max} = \sigma_{IS}/\rho_{IS} = -1$$

If the situation were $I + S \rightleftharpoons IS$ and modulation of the scalar interaction were the sole means of relaxation for I, then saturating S would result in $f = -1$, or a complete loss of the I signal. We will find later that the same effect would be obtained upon saturating S if I and S represented the same nucleus in different sites between which it was exchanging.

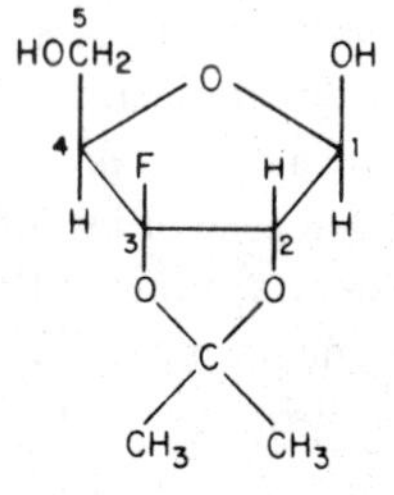

Fig. 7-1   3-Fluoro-2,3-isopropylidene-D-ribose.

A hypothetical experiment will serve to illustrate the type of situation in which scalar relaxation may affect the outcome of an NOE experiment even though neither the observed nor irradiated spins are themselves exchanging. Suppose NOE experiments were to be done on the various spins in 3-fluoro-2,3-isopropylidene-D-ribose (see Fig. 7-1) to aid in establishing its conformation in solution. If the experiments are done in water or other solvent with exchangeable hydrogen, the hydroxyl protons at carbons 5 and 1 could undergo exchange with the solvent protons. The scalar coupling of H–1 and the two H–5 protons to their respective hydroxyl protons could thus become an important relaxation mechanism for all five of the protons involved. In ignorance of the presence of exchange or the scalar coupling, the low enhancement found when H–2 was irradiated and H–1 observed, or the fluorine irradiated and H–4 observed, would be interpreted as an unexpectedly large separation between these spins, or possibly a third spin interaction as discussed in Chapter 3. In either case, error could be introduced into the interpretation.

It may be helpful in deciding whether or not scalar relaxation may be important to note that the J splitting must be at least partially collapsed in the high-resolution NMR spectrum of the sample before exchange can be fast enough to effect relaxation.

The last group of terms found in the Hamiltonian, the dipole–dipole terms, also contain x and y components, so exchange modulation of these interactions is a potential source of relaxation. It does not, however, prove to be an effective mechanism in liquids. This difference in the effect of the scalar and dipolar terms arises from a difference in their dependence on the orientation of the interacting nuclei in space. A scalar interaction between two nuclei is independent of their relative orientation.† The dipolar interaction is a tensor interaction $(\mathbf{I} \cdot \mathbf{A} \cdot \mathbf{S})$ which does depend upon the relative disposition of the spins in space—in particular, upon the angle between the direction of H_0 and a line through the interacting nuclei. Thus the thermal motions of the molecules in the liquid will modulate the dipolar but not the scalar interactions. This means that scalar interactions require exchange modulation‡ to become effective relaxation mechanisms, whereas the exchange modulation of dipolar terms must compete with the effectiveness of modulation by molecular motions. Motional effects will dominate until $\tau_{\mathrm{exch}} \geqslant \tau_{\mathrm{c}}$, where τ_{c} is the correlation time for molecular rotation. However, when exchange is that rapid we are really speaking of a diffusion controlled interaction and it is perhaps better to treat it as a case of intermolecular dipole–dipole relaxation (see Section 2,E) between species A and B than to treat it as relaxation in the complex AB with exchange effects included. The only region in which this latter treatment is really appropriate is in a narrow range of rates around $\tau_{\mathrm{exch}} = \tau_{\mathrm{c}}$.

In the case of nuclei with nonzero quadrupole moments, the Hamiltonian also contains terms corresponding to the interaction of the nucleus with local electric fields (see Section 2,B). Exchange modulation of the electric environ-

---

† It was, in fact, the lack of dependence of the splittings in the spectrum upon molecular orientation which originally led to postulating the scalar form for this interaction (*4*).

‡ Scalar relaxation may also occur if one of the coupled spins has a very short relaxation time as discussed in Section 2, G. This will generally be the case only if one of the nuclei has a quadrupole moment or is an unpaired electron.

ment can produce relaxation, but the quadrupole terms depend on the orientation of the molecule in space and $\tau_{exch} \geqslant \tau_c$ is a condition on the effectiveness of the exchange modulation (5), just as it is in the dipole–dipole case.

B. Direct Effects of Chemical Exchange: The Transfer of Magnetization between Sites

Exchange induced relaxation, as discussed in the preceding section, may be important in any Overhauser experiment on a system containing exchanging spins. The direct effects which will be discussed in this section will be most important when the observed or irradiated line is a resonance of the exchanging spin.

The direct effects are a result of the physical transfer of nuclei between sites and result in the magnetizations in all sites connected by exchange being coupled. The way in which this occurs can be seen by considering a spin-$\frac{1}{2}$ nucleus exchanging between two chemically shifted sites, A and B. We will assume that the exchange is slow enough so that separate resonances are observed for the two sites, and that the sites are equally populated. The energy levels and transition probabilities for this system are shown in Fig. 7-2. P_i is

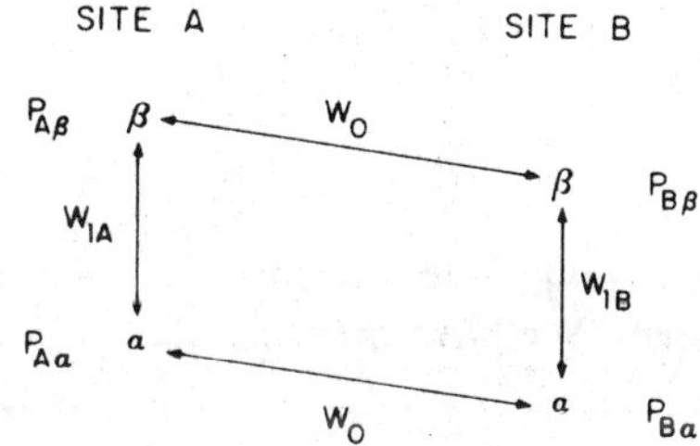

Fig. 7-2 The energy level diagram for a spin-$\frac{1}{2}$ nucleus exchanging between two sites. W_{1A} and W_{1B} are the total probabilities (by all mechanisms) for spin transitions in site A and in site B, respectively. W_0 is the probability for transfer of the spin between site A and site B.

the population of site i, W_{ij} the probability of a transition from state j to state i, and a superscript o will be used to denote an equilibrium value of a population. The intensity of the A resonance is proportional to $P_{A\alpha} - P_{A\beta}$ and the intensity of the B resonance to $P_{B\alpha} - P_{B\beta}$. Thus the equation of motion for the line intensities may be obtained from the master equation for populations [Eq. (1.9)]

$$dP_i/dt = \sum_j W_{ij}[(P_j - P_j^{\circ}) - (P_i - P_i^{\circ})] \tag{7.3}$$

Using the transition probabilities and populations from Fig. 7-2 in Eq. (7.3)

$$dP_{A\alpha}/dt = -(W_{1A} + W_0)(P_{A\alpha} - P_{A\alpha}^{\circ}) + W_{1A}(P_{A\beta} - P_{A\beta}^{\circ}) + W_0(P_{B\alpha} - P_{B\alpha}^{\circ})$$

$$dP_{A\beta}/dt = -(W_{1A} + W_0)(P_{A\beta} - P_{A\beta}^{\circ}) + W_{1A}(P_{A\alpha} - P_{A\alpha}^{\circ}) + W_0(P_{B\beta} - P_{B\beta}^{\circ})$$

Taking the difference between these equations we find

$$d(P_{A\alpha} - P_{A\beta})/dt = -(2W_{1A} + W_0)[(P_{A\alpha} - P_{A\alpha}^o) - (P_{A\beta} - P_{A\beta}^o)]$$
$$+ W_0[(P_{B\alpha} - P_{B\alpha}^o) - (P_{B\beta} - P_{B\beta}^o)]$$

and an analogous equation for $P_{B\alpha} - P_{B\beta}$. These may be rewritten in terms of $\langle \mathbf{I}_{zA}(t) \rangle$ and $\langle \mathbf{I}_{zB}(t) \rangle$ as

$$d\langle \mathbf{I}_{zA}(t) \rangle/dt = -R_A(\langle \mathbf{I}_{zA} \rangle - I_0) - \sigma_{AB}(\langle \mathbf{I}_{zB}(t) \rangle - I_0) \qquad (7.4)$$

$$d\langle \mathbf{I}_{zB}(t) \rangle/dt = -R_B(\langle \mathbf{I}_{zB}(t) \rangle - I_0) - \sigma_{BA}(\langle \mathbf{I}_{zA}(t) \rangle - I_0) \qquad (7.5)$$

where $R_A = 2W_{1A} + W_0$, $R_B = 2W_{1B} + W_0$, $I_0 = I_{0A} = I_{0B}$, and $\sigma_{AB} = \sigma_{BA} = -W_0$. Table 7-1 shows the relations between the notation used in Eqs. (7.4) and other notations common in the literature.

W_0 was defined as the probability per unit time of a nucleus transferring from site A to site B. W_0 can thus be identified with the first-order or pseudo-first-order rate constant for the exchange process by noting that

$$W_0 = \frac{\text{number of nuclei transferring per unit time}}{\text{total number of nuclei in site } A}$$

$$= \frac{k[A]V}{[A]V} = k$$

where $[A]$ is the concentration of nuclei in sites A, and V is the volume of the system. We have then that

$$\sigma_{AB} = -W_0 = -k \qquad (7.6)$$

If we use (7.6) to calculate the enhancement of A when B is saturated, we find

$$\langle \mathbf{I}_{zA} \rangle = I_0 \left(1 - \frac{k}{2W_{1A} + k} \right)$$

For $k \gg 2W_{1A}$, the fractional enhancement is $f_A(B) = -1$, or a complete loss of signal at A due to the saturation of B. Physically $k \gg 2W_{1A}$ means that the exchange of nuclei between A and B is much more rapid than interconversion of α and β spins by the relaxation processes operating at site A. This very rapid mixing of the populations of $A\alpha$ and $B\alpha$, and of $A\beta$ and $B\beta$, results in the populations of $A\alpha$ and $A\beta$ being equalized when $B\alpha$ and $B\beta$ are forced to be equal. Equalization of populations means loss of signal, or saturation of the line.

TABLE 7-1

Summary of Notations Found in the Literature

Reference, or description of alternative notation	Variable in present notation			
	R_A or $\rho_A{}^a$	σ_{AB}	R_B or $\rho_B{}^a$	σ_{BA}
Transition probabilities, simple two-site case	$W_0 + 2W_{1A}$	$-W_0$	$W_0 + 2W_{1B}$	$-W_0$
Transition probabilities, more general cases[b]	$\Sigma(W_{0A} + 2W_{1A} + W_{2A})$	$\Sigma(W_{2A} - W_{0A})$	$\Sigma(W_{0B} + 2W_{1B} + W_{2B})$	$\Sigma(W_{2B} - W_{0B})$
6–8	$\dfrac{1}{\tau_{1A}} \equiv \dfrac{1}{T_{1A}} + \dfrac{1}{\tau_A}$	$-\dfrac{1}{\tau_B}$	$\dfrac{1}{\tau_{1B}} \equiv \dfrac{1}{T_{1B}} + \dfrac{1}{\tau_B}$	$-\dfrac{1}{\tau_A}$
9, 10	ρ_A	σ_A	ρ_B	σ_B
3 (p. 295)	$(T_1^{\mathrm{II}})^{-1}$	$(T_1^{\mathrm{IS}})^{-1}$	$(T_1^{\mathrm{SS}})^{-1}$	$(T_1^{\mathrm{SI}})^{-1}$
11, 12	$[T_1(\mathrm{I})]^{-1}$	$[T_1(\mathrm{IS})]^{-1}$	$[T_1(\mathrm{S})]^{-1}$	$[T_1(\mathrm{SI})]^{-1}$

[a] When several relaxation paths are available to a spin, the ρ's are replaced by $R_A = \Sigma_i \rho_{Ai}$ in the equation of motion for the spin system.

[b] The sums run over all levels directly accessible to a spin in energy level A.

Another consequence of Eq. (7.6) is that Overhauser experiments may be used to measure the rate of exchange under favorable circumstances. The experiments described in Chapter 6 can be used to evaluate the R's and σ's for any system obeying the equations of motion (6.1), and Eqs. (7.4) are just special cases of (6.1). There are, however, two requirements that must be met in order for this to be useful in studying exchange.

(1) $k < \Delta\omega$ where $\Delta\omega$ is the chemical shift (in radians per second, i.e., $2\pi\delta$) between lines in the spectrum corresponding to the different exchange sites. If this condition is not satisfied, the lines will overlap or coincide and NOE experiments between them will not be possible.

(2) W_1 and k $(= W_0)$ must be of similar magnitude in order that information on the exchange rate can be derived from any Overhauser experiment. If exchange were very slow, so that $k \ll W_1$, we would have

$$f_A(B) = -k/(2W_{1A} + k) \approx -k/2W_{1A} \approx 0$$

and, similarly, $f_B(A) = 0$. In addition to no Overhauser effect, any measurement of relaxation times would determine only W_{1A}^{-1} and W_{1B}^{-1}. A similar conclusion is reached for $k \gg W_{1A}$ where $f_A(B) = f_B(A) = -1$. The experiment gives the same result no matter what the exact value of k is and thus k cannot be derived from it.

We can estimate the range of values of the ratio W_{1A}/k over which Overhauser experiments will be useful in evaluating k. We will assume that the smallest measurable signal is 3% of I_0 and the smallest measurable change in $\langle \mathbf{I}_{zA} \rangle$ is also 3% of I_0. The relation $f = -k/(2W_{1A} + k)$ can be rearranged to $2W_{1A}/k = -1 - 1/f$. The limits set by detectability are then $-0.03 \geqslant f \geqslant -0.97$ and so we have $64.7 \geqslant k/W_{1A} \geqslant 0.06$. While this is only a rough estimate of the limits encountered in practice, it is apparent that the NOE method applies only to a rather narrow range of rates in the slow exchange region. It is, however, a range which is not readily accessible by other techniques.

The extension of this discussion to the general case of exchange between N sites $(i, j, k, ..., N)$ is very straightforward. Equations (6.1) remain valid with $\rho_{ij,\,\text{exch}} = 2W_{1A} + k_{ij}$ and $\sigma_{ij} = -k_{ij}$. As there is always the possibility that spins in site i may be coupled with those in j by dipole–dipole or scalar interaction as well as by exchange, Eq. (6.1) must be rewritten

$$d\langle \mathbf{I}_{zi} \rangle/dt = -R_i(\langle \mathbf{I}_{zi} \rangle - I_{0i}) - \sum_{j \neq 1} S_i(j)(\langle \mathbf{I}_{zj} \rangle - I_{0j}) \qquad (7.7)$$

where $R_i = \sum_m \sum_j \rho_{ij}$ and $S_i(j) = \sum_m \sigma_{ij}$. The sum over j is a sum over all the spins that interact with i, and the sum over m runs over all the relaxation mechanisms (scalar, dipolar, exchange) that couple spins i and j.

We have already mentioned that the multiple resonance technique offers the possibility of studying kinetics in a region of rate constants not otherwise accessible by NMR methods. Additional advantages arise in multiple-site problems. The first of these is that all the processes for which R_i/k_{ij} fall in the proper range can be studied on a single sample which is at chemical equilibrium.†

A second advantage is that the Overhauser technique can provide qualitative information on the sequence of steps in a reaction. In particular, the equality of the results from two different types of Overhauser experiments on two of the sites will indicate that direct exchange between those sites is not an important process.

Forsén and Hoffman (6) first suggested this application of multiple resonance to sorting out reaction paths, but discussed using a transient experiment (Experiment 9 of Chapter 6) in place of the steady state NOE experiments. Our notation is related to theirs by

$$R_k = \frac{1}{T_{1k}} + \sum_{v}{}' \lambda_{kv} = \frac{1}{\tau_{1k}}$$

and $+ k_{vk} = \lambda_{vk}$. As transient experiments are more difficult to perform than steady-state experiments, the procedure suggested here should prove more useful.

Suppose it is known that a spin may move from site A to site B, but it is not known whether the spin passes directly between these sites or if intermediate sites are necessary. It is assumed that the A and B resonances are isolated from each other and from other lines in the spectrum. Three multiple resonance experiments are then performed.

(1) The NOE is observed for line A while saturating all other lines in the spectrum. The resulting enhancement is given by

$$f_A(\text{all other spins}) = \left(\sum_{j \neq A} S_A(j)\, I_{0j} \right) \Big/ I_{0A}\, R_A \tag{7.8}$$

(2) Experiment 1 is repeated, but now observing B while saturating all other lines. The enhancement will be

$$f_B(\text{all other spins}) = \left(\sum_{j \neq B} S_B(j)\, I_{0j} \right) \Big/ I_{0B}\, R_B \tag{7.9}$$

---

† Although removed from the immediate subject, it is of interest to note that in general, the equilibrium constant for a reaction is the ratio of the forward and reverse rate constants *only* when those constants are measured at equilibrium. At least one case appears in the literature (*13, 14*) where the ratio of forward and reverse rate constants does not give the correct equilibrium constant when the rates are measured far from equilibrium.

(3) The NOE on A and B is observed while saturating all other resonances due to the exchanging spin. This is Experiment 10 of Chapter 6. The enhancements are given by

$$f_B(\text{all spins } M, M \neq A, B) = (\langle \mathbf{I}_{zA} \rangle - I_{0A}) S_B(A)/I_{0B} R_B$$

$$+ \left(\sum_{j \neq A,B} S_B(j) I_{0j} \right) \Big/ I_{0B} R_B \qquad (7.10)$$

$$f_A(\text{all spins } M, M \neq A, B) = (\langle \mathbf{I}_{zB} \rangle - I_{0B}) S_A(B)/I_{0A} R_A$$

$$+ \left(\sum_{j \neq A,B} S_A(j) I_{0j} \right) \Big/ I_{0A} R_A \qquad (7.11)$$

Comparing Eqs. (7.8)–(7.11) we find that the enhancements in the two experiments on A (1 and 3) will be the same if

$$S_A(B)(\langle \mathbf{I}_{zB} \rangle - I_{0B}) \ll \sum_{j \neq A,B} S_A(j) I_{0j}$$

Similarly, the results of experiments on B (2 and 3) will be the same if

$$S_B(A)(\langle \mathbf{I}_{zA} \rangle - I_{0A}) \ll \sum_{j \neq A,B} S_B(j) I_{0j}$$

Assuming that reasonably large enhancements of A and B are observed, the inequalities imply that the only way we can have

$$f_A(\text{all } M, M \neq A) = f_A(\text{all } M, M \neq A, B)$$

and

$$f_B(\text{all } M, M \neq B) = f_B(\text{all } M, M \neq A, B)$$

will be for $S_B(A)$ and $S_A(B)$ themselves to be small. This means that direct exchange between A and B is not important, as $|S_A(B)| \geqslant k_{AB}$ and $|S_B(A)| \geqslant k_{BA}$, the equality holding if exchange is the only process coupling sites A and B. As with other Overhauser experiments, these results should be interpreted with caution. Where possible, the preferred method is actually to evaluate all the rate constants and determine whether k_{AB} and k_{BA} are smaller than the other k_{ij} by direct comparison.

C. A Review of Applications of NOE to Exchanging Systems: Intermolecular Exchange

The results of Feeney and Heinrich (*15*) nicely illustrate the basic direct effect of exchange on an NOE experiment. These studies were conducted on phenolic compounds dissolved in deuterochloroform containing a trace of water. The protons of the water undergo slow, reversible exchange with the phenolic protons. As the water resonance is well separated from the resonance of the phenolic protons ($\delta \geqslant 2$ ppm) it is very straightforward to irradiate the water resonance and observe the effect this has on the phenolic proton resonance. Figure 7-3 shows the results obtained for 2,4,5-trimethylphenol.

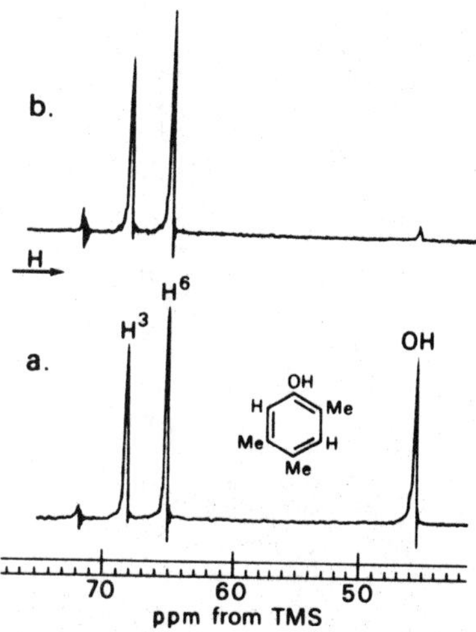

Fig. 7-3 Proton NMR spectra of 2,4,5-trimethylphenol in solution in CDCl$_3$. The spectra were recorded at 100 MHz with the spectrometer operating in the frequency-sweep mode. a, Normal spectrum. b, Spectrum with strong irradiation of the DOH proton resonance (*15*).

Saturation of the water resonance (not shown in the figure) results in complete saturation of the phenolic proton resonance. The results they obtained for other systems are summarized in Table 7-2 and, except for 2-acetyl-3-methoxyphenol, they directly parallel those obtained for trimethoxyphenol. In 2-acetyl-3-methoxyphenol, the decrease in intensity of the phenolic peak was only 15% when the water resonance was saturated, presumably because the internal hydrogen bonding (see structure in Table 7-2) slows the exchange to a point where relaxation in the sites is more rapid than exchange between the sites ($W_{1A} > W_0 = k$). Feeney and Heinrich suggest using this technique to identify the lines in a spectrum which are capable of exchanging with water protons. They point out, however, that in aliphatic alcohols and carboxylic acids the exchange is so rapid that only a single resonance is observed for all exchanging protons ($k \gg |\omega_A - \omega_{H_2O}|$) so that the method could not be used in these cases.

Fung and Stolow (*16*) performed NOE experiments on exchanging protons in two systems. The first system was a CS$_2$ solution that was 0.60 molal in diphenylmethanol and 0.58 molal in *t*-butanol. The study was done at 25°C and the normal NMR spectrum of their solution at this temperature is shown

TABLE 7-2

SUMMARY OF NEGATIVE OVERHAUSER EFFECTS OBSERVED BY FEENEY
AND HEINRICH[a] IN SYSTEMS WITH EXCHANGING PROTONS

Compound	Chemical shift of irradiated DOH line δ(ppm)[b]	Decrease in intensity of the phenolic OH resonance (%)
(A)	1.60	100
(B)	1.53	100
(C)	1.20	100
(D)	1.30	100
(E)	1.52	100

TABLE 7-2 (Continued)

Compound	Chemical shift of irradiated DOH line δ(ppm)[b]	Decrease in intensity of the phenolic OH resonance (%)
(F) phenol with OH, Cl (×3)	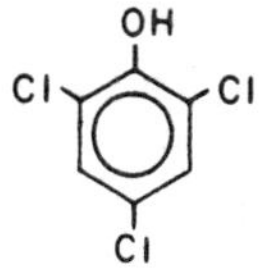 1.50	100
(G) phenol with OH, CH₃ (×3)	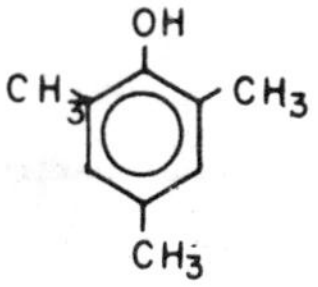 1.73	100
(H) benzene ring with OH···O=C–CH₃ and OCH₃	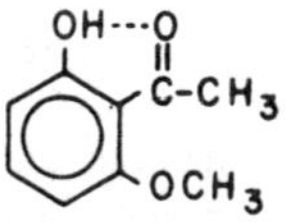 1.75	15
(I) naphthalene with OH	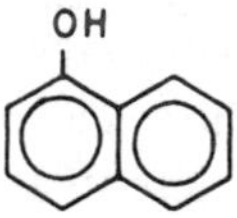 1.65	100
(J) benzene ring with OH, HOOC, Cl	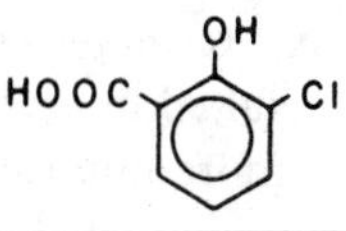 1.62	100[c]

[a] See Feeney and Heinrich (*15*).
[b] δ is measured in parts per million from TMS.
[c] The $-CO_2H$ resonance also showed a partial loss of intensity.

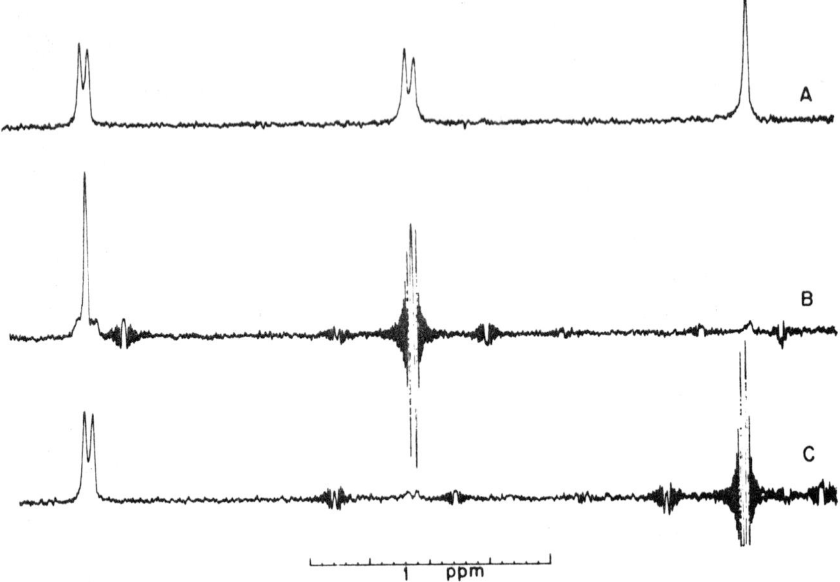

Fig. 7-4 Proton NMR spectra of diphenylmethanol (0.60 molal) and *t*-butanol (0.58 molal) in CS$_2$ at 25°C and 100 MHz. A, normal spectrum of the solution. B, The spectrum with a strong rf field at the frequency of the –OH proton of diphenylmethanol. Note the decoupling of the –CH proton of diphenylmethanol and the loss of signal from the –OH proton of *t*-butanol. C, The spectrum with a strong rf field at the frequency of the *t*-butanol –OH proton. Note that the loss of signal from the diphenylmethanol –OH proton occurs without decoupling of the methine proton. [Fung and Stolow (*16*).]

in Fig. 7-4A. The C–H and –COH proton resonances of the diphenylmethanol are doublets due to their mutual *J* coupling. The second system studied was a carbon disulfide solution that was 0.85 molal in benzyl alcohol and 0.74 molal in *t*-butanol. This solution was studied at 20°C and its spectrum is given in Fig. 7-5A. In this system the COH proton resonance is a triplet due to *J* coupling with the benzylic protons, and the benzylic protons appear as a doublet. The hydroxyl resonance of the *t*-butanol is, of course, a singlet in both systems.

Fung and Stolow found that irradiating the hydroxyl resonance of diphenyl-methanol decoupled the methine proton and also resulted in a complete loss of signal from the *t*-butanol hydroxyl proton. Even more interesting was the observation that saturating the resonance of the butanol hydroxyl proton caused a complete loss of signal from the hydroxyl proton of the diphenyl-methanol, but did *not* collapse the methine doublet. This behavior was also observed in the benzyl-alcohol–*t*-butanol system.

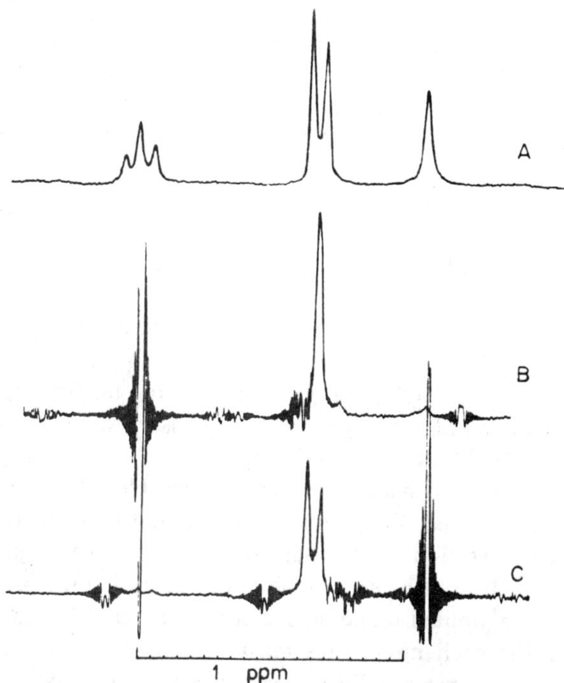

Fig. 7-5 Proton NMR spectra of benzyl alcohol (0.85 molal) and *t*-butanol (0.74 molal) in CS$_2$ at 20°C and 100 MHz. A, Normal spectrum of the solution. B, The spectrum when a strong saturating rf field is applied at the frequency of the benzyl alcohol –OH resonance. Note the decoupling of the benzyl protons and the loss of signal from the *t*-butanol–OH resonance. C, The spectrum when a strong rf is applied at the frequency of the *t*-butanol–OH resonance. Note that the loss of signal from the benzyl-alcohol–OH resonance occurs without decoupling of the benzyl protons. [Fung and Stolow (*16*).]

The reason that decoupling is not observed in the latter case lies in the difference between saturation and decoupling. First, recall that

$$J\mathbf{I}\cdot\mathbf{S} = J(\mathbf{I}_x\mathbf{S}_x + \mathbf{I}_y\mathbf{S}_y + \mathbf{I}_z\mathbf{S}_z) \tag{7.12}$$

We will assume that it is spin I which is irradiated. Saturation of the line occurs when the rate of transitions induced by the rf field far exceeds the rate at which relaxation occurs so that $\langle\mathbf{I}_z\rangle = 0$; this will occur when $\gamma_{\mathrm{I}}^2 H_2^2 T_2 T_1 \gg 1$, where H_2 is the rf field strength and T_1 and T_2 are the longitudinal and transverse relaxation times of I, respectively. The fact that $\langle\mathbf{I}_z\rangle$ is zero does not, however, require that either $\langle\mathbf{I}_z\mathbf{S}_z\rangle$ or $J\mathbf{I}\cdot\mathbf{S}$ be zero. Thus saturation of I is not sufficient for decoupling I and S.

In order to obtain rf decoupling the irradiating field must be strong enough to cause I to oscillate at a frequency much greater than J. These oscillations result in averaging $\mathbf{I} \cdot \mathbf{S}$ to zero and will generally require a field strength satisfying $\gamma_1 H_2/2 \gg J$, which is usually more restrictive than the requirement for saturation. "Exchange decoupling," on the other hand, occurs when $k \gg J$ and is also a result of averaging $J\mathbf{I} \cdot \mathbf{S}$ to zero. The behavior in partially decoupled cases will, however, be much different in exchange decoupled cases than in rf decoupled cases because the exchange process averages by a random modulation of the scalar interaction whereas the rf averaging results from a coherent oscillation. Hoffman and Forsén (*17*) have given a rather good discussion of this problem.

We can now see why decoupling did not occur in the examples presented by Fung and Stolow. Exchange is too slow to decouple the spins and the strong rf field applied at the butyl hydroxyl signal is much too far from resonance to drive the magnetization of the downfield hydroxyls in rapid oscillation. The coupling remains although the signal disappears.

Because the coupling constants remain observable while exchangeable proton signals are removed, Fung and Stolow suggest that this technique be used in place of deuteration to eliminate signals in the spectrum due to exchangeable protons. It must be kept in mind, however, that the procedure will only work for a very limited range of rate constants, and that the relaxation time in each of the exchanging sites must be long ($T_1 > k^{-1}$). Fung has also discussed the extension of this procedure to tightly coupled systems (*18–20*).

Forsén and Hoffman (*6–8*) were the first to investigate the potential of Overhauser experiments for studying the exchange process itself. Their second paper (*8*) on this subject dealt with the exchange of hydroxyl protons between salicylaldehyde and 2-hydroxyacetophenone in carbon disulfide containing a trace of acetic acid. Experiments of type 1, 3, 6, and 7 of Chapter 6 were done on this system in order to illustrate the principles of the method and check on the correctness of the theory. Since experiments of type 1 are most easily interpreted, the dynamic parameters of the spin system (R's and σ's) were determined from experiments of this type. The parameter values determined in this way were then used to calculate the decay curves that would be expected in Experiments 3, 6, and 7. The correctness of the theory was checked by comparing these calculated results to the actual experimental results.

As discussed in detail in Chapter 6, Experiment 1 involves suddenly turning on the saturating field at resonance B and monitoring the decay of resonance A to a new steady-state value; Experiment 3 is then to observe the return of A to equilibrium when the saturating field is removed from B. The decay curves obtained when these two experiments were performed on the salicylaldehyde–2-hydroxyacetophenone system are shown in Fig. 7-6. The plots of $\log[M_z - \tau_{1A} M_z^\circ/T_{1A}]$ used in evaluating R_A and R_B are shown in Fig. 7-7;

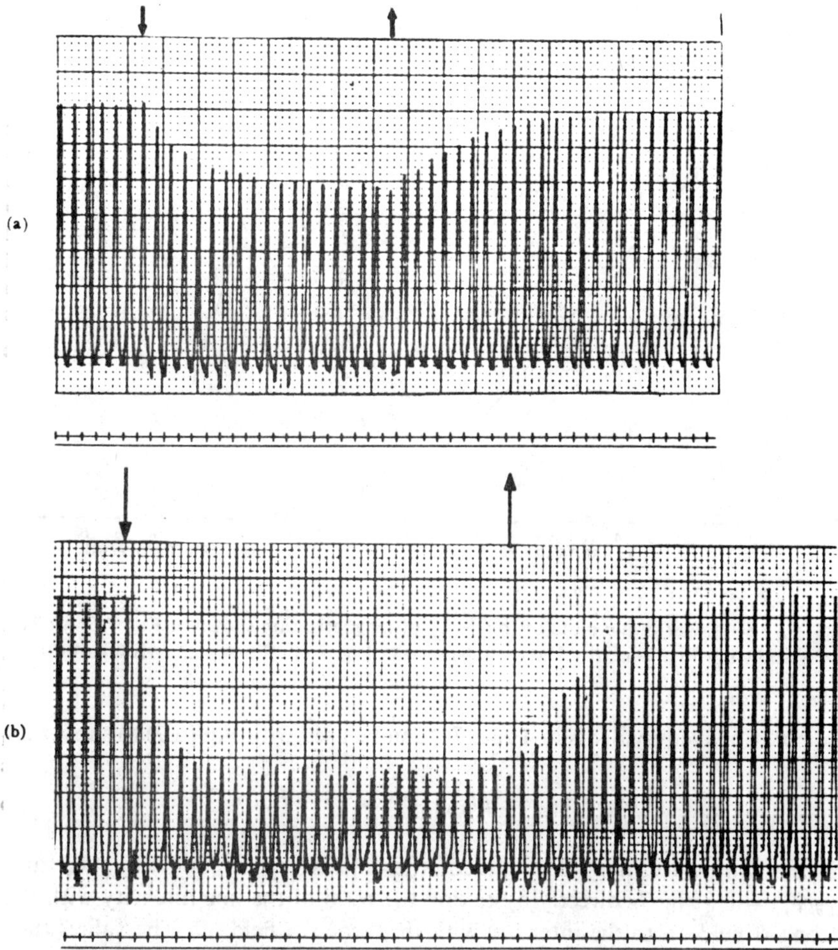

(a)

(b)

Fig. 7-6 Transient NMR experiments on a solution of salicylaldehyde (A) and 2-hydroxy-acetophenone (B) in CS_2. The arrows pointing downward ($\downarrow$) indicate the moment at which the saturating rf field is turned on (Experiment 1 of Chapter 6) and the arrows pointing upward ($\uparrow$) indicate the moment the saturating field is turned off (Experiment 3 of Chapter 6). The markers in the lower part of the figures are second intervals. (a) The decay of the hydroxyl resonance of A to a new steady-state intensity upon suddenly saturating the hydroxyl resonance of B, and the recovery of A upon removing the saturating field from resonance B. (b) The analogous decay and recovery of the B hydroxyl resonance. [Forsén and Hoffman (8).]

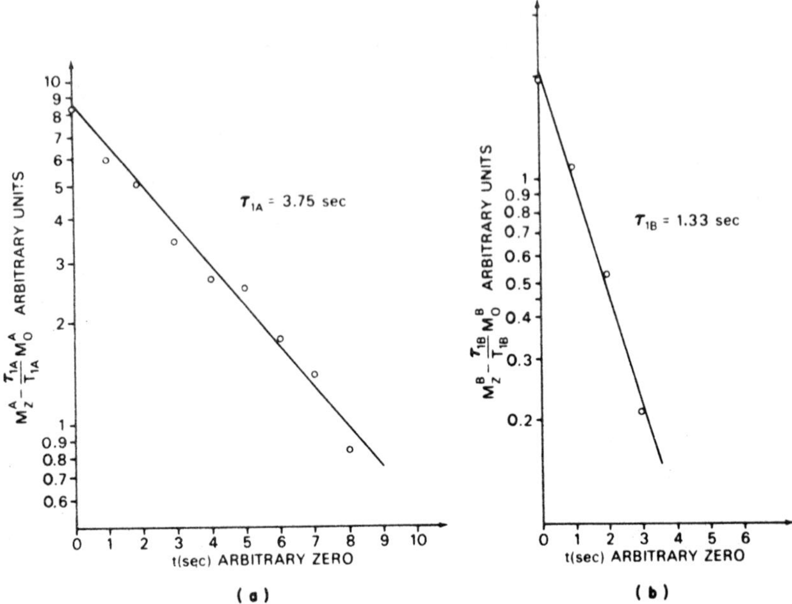

Fig. 7-7   Semilogarithmic plots of the decay of magnetization shown in Fig. 7-6. (a) The decay shown in Fig. 7-6a. (b) The decay shown in Fig. 7-6b. [Forsén and Hoffman (8).]

the notational equivalents in Table 7-1 can be used with Eq. (6.7) to show that

$$[M_z - \tau_{1A} M_z^\circ / T_{1A}] \propto [\langle \mathbf{I}_{zA} \rangle - I_{zA}(\infty)]$$

Forsén and Hoffman determined the ratios $\tau_{1A}/T_{1A} = 1 + \sigma_{BA}/R_A$ and $\tau_{1B}/T_{1B} = 1 + \sigma_{AB}/R_B$ directly by taking the ratio of the line intensity with the strong rf field on to the intensity with the strong rf field off. The values they obtained for the constants are given in Table 7-3. Table 7-3 also contains results obtained (8) on a solution that was 50% carbon disulfide and 50% an equimolar mixture of *t*-butanol and 2-hydroxyacetophenone. The precisions given in the table are mean deviations calculated from the data on repetitions of these experiments given by Forsén and Hoffman (6, 7). The precision in determining $1 + \sigma/R$ is quite good ($\leqslant 4\%$ in all four cases) and thus contributes only about 1% to the uncertainty in k. A check on the accuracy is provided by noting that a two site system in a state of chemical equilibrium necessarily satisfies the condition $k_{AB}[B] = k_{BA}[A]$ so that

$$\frac{k_{AB}}{k_{BA}} = \frac{[A]}{[B]} = \frac{\text{integral of resonance } A}{\text{integral of resonance } B}$$

TABLE 7-3

RATE CONSTANTS FOR CHEMICAL EXCHANGE AND SPIN RELAXATION IN
TWO SYSTEMS STUDIED BY FORSÉN AND HOFFMAN

Rate constant[a]	Original notation[b,c]	Salicylaldehyde (A) + 2-hydroxy-acetophenone (B)[b,d] (sec^{-1})	t-butanol (A) + 2-hydroxy-acetophenone (B)[c,d] (sec^{-1})
$-\sigma_{BA} = k_{BA}$	τ_A^{-1}	0.0869, $\pm 5\%$	0.476, $\pm 6\%$
$-\sigma_{AB} = k_{AB}$	τ_B^{-1}	0.455, $\pm 3\%$	0.435, $\pm 13\%$
R_A	τ_{1A}^{-1}	0.270, $\pm 4\%$	0.752, $\pm 5\%$
R_B	τ_{1B}^{-1}	0.730, $\pm 2\%$	0.675, $\pm 12\%$

[a] k_{AB} = rate constant for the transfer of spins from site B to site A.
[b] Data from Forsén and Hoffman (8).
[c] Data from Forsén and Hoffman (7).
[d] The precisions listed in the table are relative mean deviations calculated from the data on repetitions of the experiments given by Forsén and Hoffman (6,7).

The ratios of the integrals can be determined with good accuracy in an independent experiment on the high-resolution spectrum. In the t-butanol system it was found that $k_{AB}/k_{BA} = 0.9$, as compared with an intensity ratio of 1.1. The values for the salicylaldehyde system were found to be $k_{AB}/k_{BA} = 5.3$ and an intensity ratio of 5.65. The discrepancy is 12% and 6%, respectively, which is quite satisfactory. The use of the data from Experiment 1 to predict the outcome of the experiments to be discussed below was quite successful, and this may also be taken as an indication of the accuracy of the results and correctness of the theory. On the basis of the data provided, no further check on the accuracy of the method is possible.

λ_1 and λ_2, the exponents characterizing the decays in Experiments of type 3, 6, and 7, can be calculated using Eq. (6.6) and the data in Table 7-3. This yields $\lambda_1 = 0.805$ sec^{-1} and $\lambda_2 = 0.196$ sec^{-1}. Then using Eqs. (6.6), (6.4), and (6.9) the decay expected in Experiment 3 becomes

$$\langle I_{zA}(t) \rangle = I_{0A} + I_{0A}(0.103e^{-0.805t} - 0.423e^{-0.196t}) \tag{7.13a}$$

$$\langle I_{zB}(t) \rangle = I_{0B} + I_{0B}(0.203e^{-0.805t} + 0.833e^{-0.196t}) \tag{7.13b}$$

The curves obtained in this experiment are shown in Fig. 7-6 and semilog plots of the data in Fig. 7-8. The first term in Eq. (7.13) decays more rapidly than the second term because it has a larger exponent. This term can only be important in the early stages of the decay and accounts for the curvature of the line near $t = 0$. The later portions of the semilog plot correspond to the decay

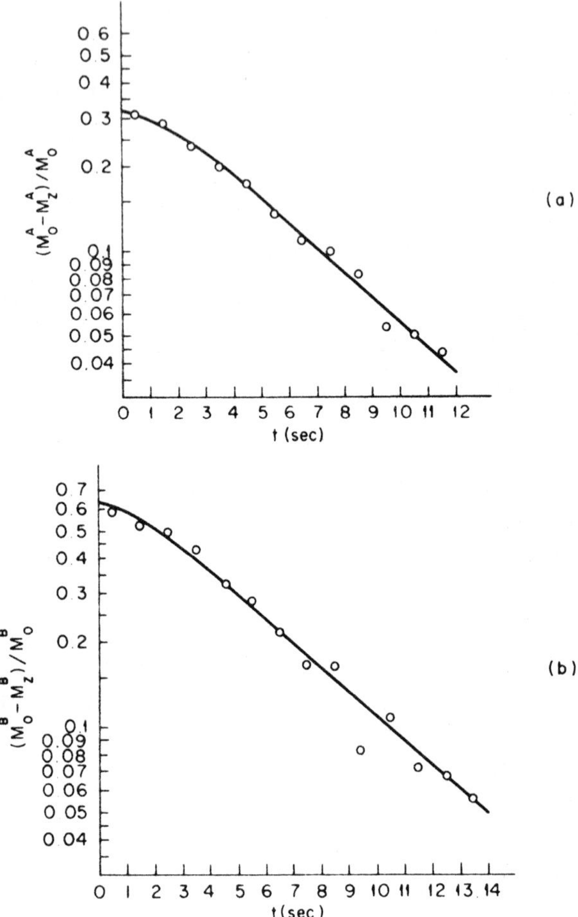

Fig. 7-8 Semilogarithmic plots of the return of the magnetization to equilibrium for the experiment shown in Fig. 7-6. (a) The recovery shown in Fig. 7-6a. (b) The recovery shown in Fig. 7-6b. The solid lines represent the theoretically predicted recoveries (see p. 143). [Forsén and Hoffman (8).]

of the second term. Similar considerations apply to the plot obtained for the B resonance decay shown in Fig. 7-8b. Note that it was necessary to introduce a time lag of 0.5 sec into the calculation in order to superimpose the theoretical curve on the experimental data as shown in Fig. 7-8. This delay would correspond to the time lag between removing the saturating field and making the first observation. Forsén and Hoffman indicated that, considering their experimental procedure, this was a reasonable value for the lag.

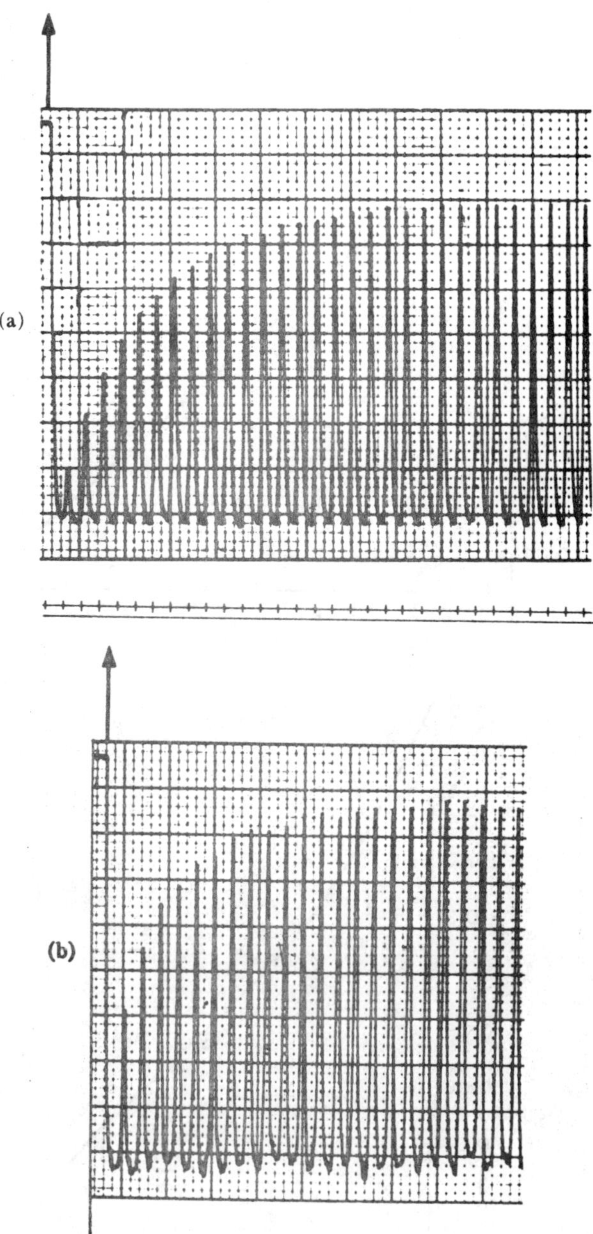

Fig. 7-9 The return of the signal intensity to equilibrium after removal of a saturating rf field from it (Experiment 6 of Chapter 6). The experiment was performed on a solution of salicylaldehyde (A) and 2-hydroxyacetophenone (B) in CS_2. The arrows ($\uparrow$) indicate the moment when the saturating field was removed. The markers along the bottom edge of the figures are second intervals. (a) The recovery of the hydroxyl resonance of A. (b) The recovery of the hydroxyl resonance of B. [Forsén and Hoffman (8).]

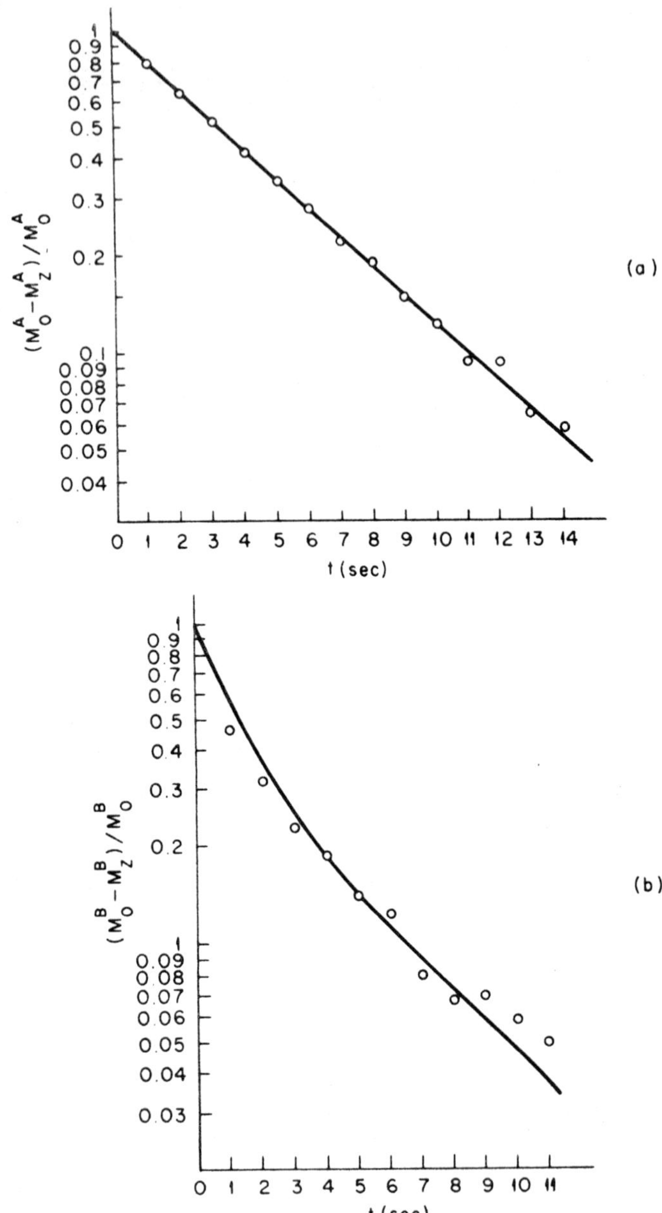

Fig. 7-10 Semilogarithmic plots of the return of the magnetization to equilibrium for the experiment shown in Fig. 7-9. (a) The recovery shown in Fig. 7-9a. (b) The recovery shown in Fig. 7-9b. The solid lines represent the theoretically predicted recoveries (see p. 149). [Forsén and Hoffman (8).]

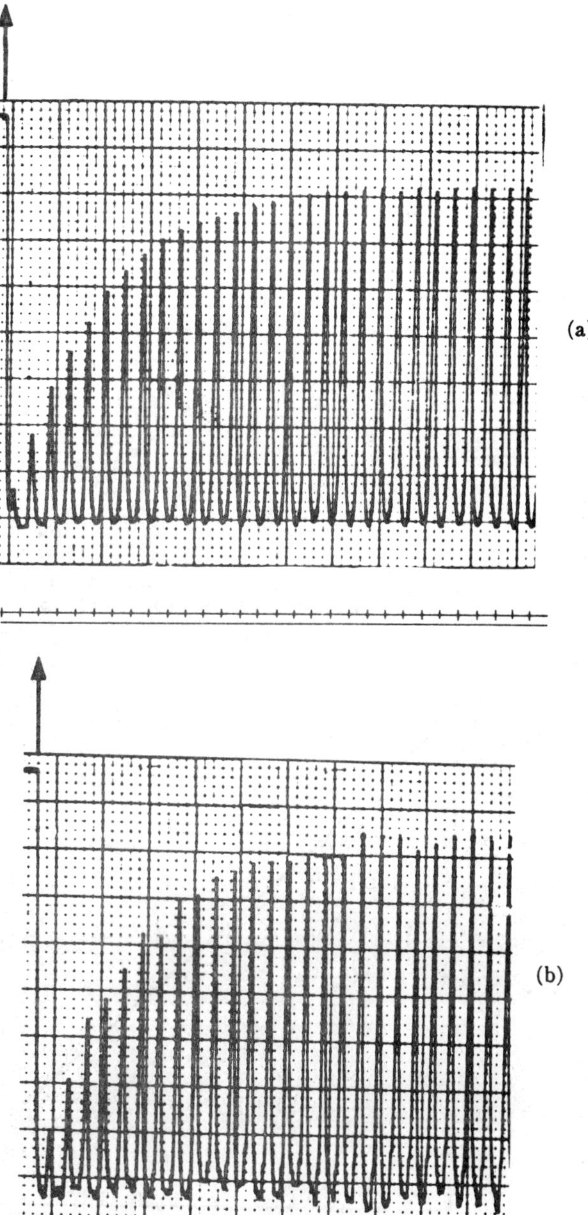

Fig. 7-11 The return of the signal intensity to equilibrium upon simultaneously removing the saturating rf fields from the hydroxyl resonances of both salicylaldehyde (*A*) and 2-hydroxyacetophenone (*B*) (Experiment 7 of Chapter 6). The arrows (↑) indicate the moment when the saturating rf fields were removed. The markers along the lower edge of the figures are second intervals. (a) The recovery of the hydroxyl signal of *A*. (b) The recovery of the hydroxyl signal of *B*. [Forsén and Hoffman (8).]

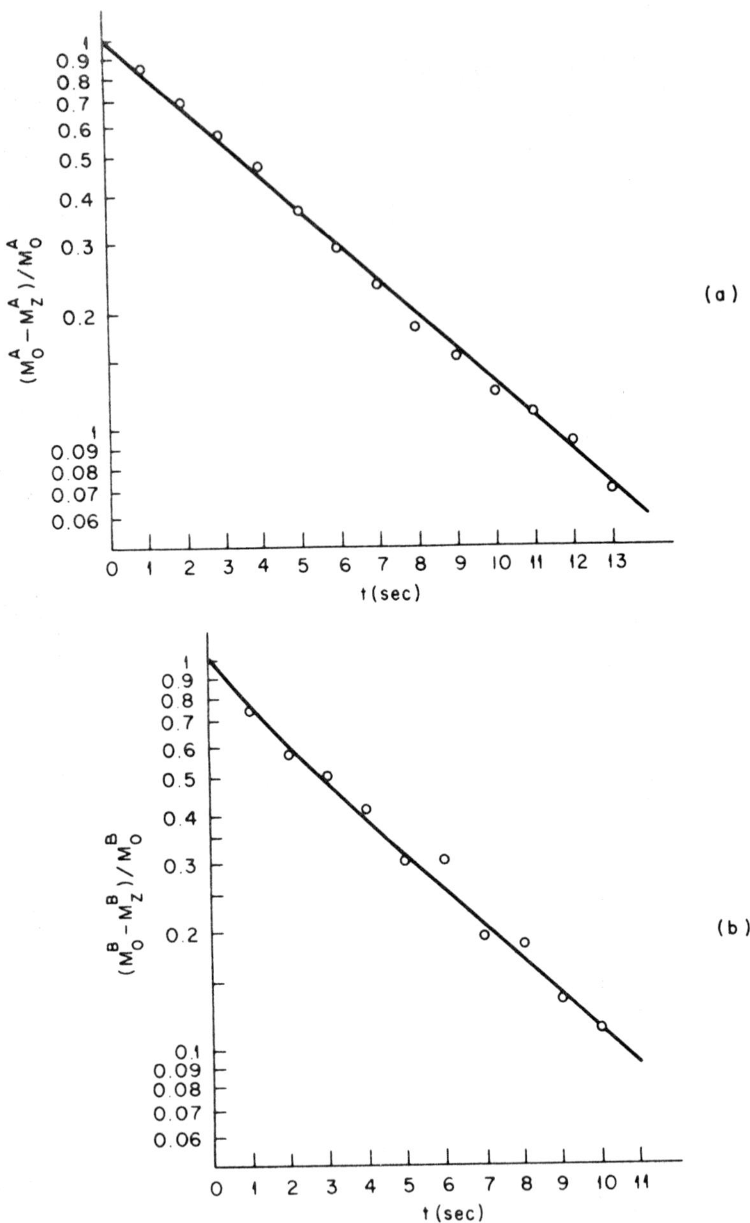

Fig. 7-12 Semilogarithmic plot of the recovery of the signal intensity for the experiment shown in Fig. 7-11. (a) The recovery shown in Fig. 7-11a. (b) The recovery shown in Fig. 7-11b. The solid lines represent the theoretically predicted recoveries (see p. 150). [Forsén and Hoffman (8).]

Experiment 6 consists of saturating one resonance and then watching its recovery when the saturating field is removed. The experimental recovery curves are shown in Fig. 7-9. Using Eqs. (6.4), (6.5), and (6.15) the values of λ_1 and λ_2 computed above, and the constants in Table 7-3, the decay curves are expected to follow the equations

$$\langle \mathbf{I}_{zA}(t) \rangle = I_{0A} - I_{0A}(0.967e^{-0.196t} + 0.033e^{-0.805t})$$

$$\langle \mathbf{I}_{zB}(t) \rangle = I_{0B} - I_{0B}(-0.365e^{-0.196t} + 0.635e^{-0.805t})$$

The solid lines in Fig. 7-10 are computed using these equations and show reasonable agreement with the experimental points. The decay of A is almost

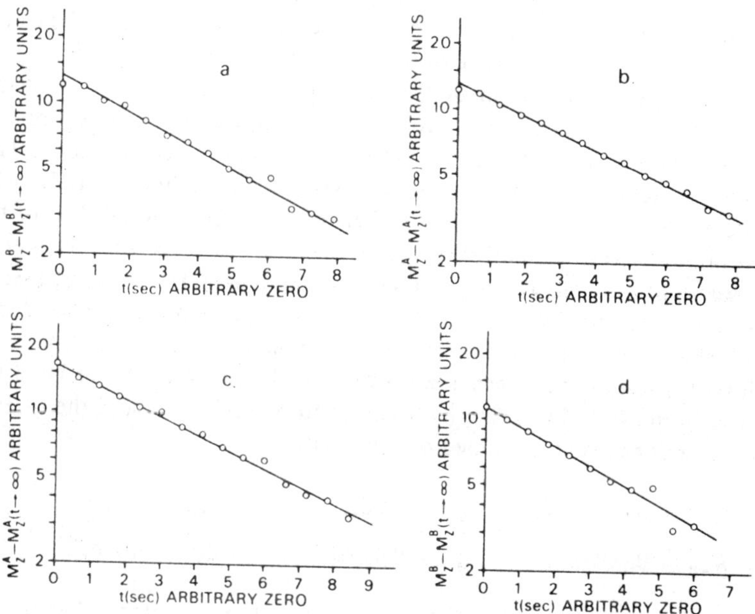

Fig. 7-13 a, Semilogarithmic plot of the decay of the –OH proton resonance of the enol form of acetylacetone when the resonances of the –CH= proton of the enol and –CH$_2$– protons of the keto form are suddenly saturated. The decay time τ_{1B} is 5.14 sec. b, Semilogarithmic plot of the decay of the olefinic –CH= proton resonance of the enol form of acetylacetone when the resonances of the –OH proton of the enol and –CH$_2$– proton of the keto form are suddenly saturated. The decay time τ_{1A} is 5.75 sec. c, Semilogarithmic plot of the decay of the –CH= proton resonance of the enol form of acetylacetone when the resonance of the –CH$_2$– protons of the keto form is suddenly saturated. The decay time is 5.61 sec. d, Semilogarithmic plot of the decay of the –OH proton resonance of the enol form of acetylacetone when the resonance of the –CH$_2$– protons of the keto form is suddenly saturated. The decay time is 4.86 sec. [Forsén and Hoffman (6).]

exponential because the coefficient of the second term is so small. The decay of the B resonance, on the contrary, shows the effects of both exponentials throughout the period of observation in this experiment.

The last experiment reported on this system was of type 7, in which both lines are saturated, the strong rf fields are removed, and the decay of the chosen line back to its equilibrium intensity is monitored. The experimental decays are shown in Fig. 7-11. The solid line is drawn according to

$$\langle \mathbf{I}_{zA}(t) \rangle = I_{0A} + I_{0A}(0.020e^{-0.805t} - 1.020e^{-0.196t})$$

$$\langle \mathbf{I}_{zB}(t) \rangle = I_{0B} + I_{0B}(0.121e^{-0.805t} + 0.879e^{-0.196t})$$

which are the equations obtained by substituting the parameters from Table 7-3 into Eqs. (6.4), (6.5), and (6.15). Semilog plots of the decay data versus time are shown in Fig. 7-12. Again, the agreement between the calculated and actual time dependence of the line intensities is quite good.

Forsén and Hoffman also made the first extension and application of this theory to multiple-site exchange problems (6). They chose to study proton exchange between three sites in the keto–enol system of acetylacetone. This system had previously been studied by Reeves and Schneider (21), who suggested that the exchange of protons between the –OH and –CH=C of the enol form occurred through the keto intermediate. The three sites of interest in the exchange problem were thus –CH=C, –OH, and –CH$_2$–. It was possible in this case to perform Experiments 8 and 10 of Chapter 6 on all spins and spin pairs, so that all dynamic parameters were completely determined. The semilog plots of the observed decays are shown in Figure 7-13. If we label the olefinic proton –CH= by A, the hydroxyl proton –OH by B, and the –CH$_2$– protons by C, the results obtained may be written:

$$k_{BA} = 0.00\,\text{sec}^{-1}, \qquad k_{CA} = 0.08\,\text{sec}^{-1}, \qquad k_{CB} = 0.06\,\text{sec}^{-1}$$

$$k_{AB} = 0.00\,\text{sec}^{-1}, \qquad k_{AC} = 0.11\,\text{sec}^{-1}, \qquad k_{BC} = 0.10\,\text{sec}^{-1}$$

The conclusion of this study was that, at least within experimental error, the direct exchange of protons between the –CH= and –OH of the enol form was negligible. Several cross checks were made on the experimental data and may be summarized as follows:

(1) In the absence of direct exchange between the A and B sites, the condition for chemical equilibrium is

$$k_{CA}/k_{AC} = k_{CB}/k_{BC} = I_{0C}/I_{0A}$$

where the fact that $I_{0A} = I_{0B}$ has also been used. The experimental results were $k_{CA}/k_{BC} = 0.6$ and $I_{0C}/I_{0A} = 0.50$ (obtained by integration of the normal high-resolution spectrum).

(2) If the only direct exchange processes are $A \rightleftharpoons C \rightleftharpoons B$, then the decay of A upon sudden saturation of C should show simple exponential behavior with the same exponent as that found when both B and C are suddenly saturated (Experiment 8), and similarly for the decay of B. That this is true can be seen in Fig. 7-13.

(3) Defining the lifetime τ_i of a nucleus at the ith site by $1/\tau_i = \sum_j k_{ji}$, in the absence of direct exchange between A and C one expects $k_{CA} = 1/\tau_A = 0.07$ sec^{-1}, $k_{CB} = 1/\tau_B = 0.07$ sec^{-1}, and $k_{AC} = k_{CA} = 1/\tau_C = 0.26$ sec^{-1}, which conditions were fulfilled.

(4) The value of $\tau_C/\tau_A = [\text{keto}]/[\text{enol}]$ obtained from the multiple resonance experiments was 0.27, in good agreement with the ratio of 0.25 found by integrating the spectrum.

The studies we have just discussed were all carried out on systems undergoing very slow exchange. Partly for this reason and partly because of the molecules involved, interpretation of the results was not complicated by the presence of exchange induced relaxation. However, in reporting on a study of nuclear spin relaxation in anhydrous liquid HF, Solomon (9) found the results to be inconsistent with the hypothesis that only dipole–dipole relaxation between H and F in a given molecule was important. The discrepancy could have been due to significant contributions from intermolecular relaxation, the existence of a second relaxation mechanism, or some combination of these. A subsequent paper by Solomon and Bloembergen (10) showed that inclusion of exchange induced scalar relaxation eliminates all discrepancies between the calculated and observed decays. This case is, in fact, quite the opposite of the ones we have just reviewed, because exchange induced relaxation is important but there are no direct effects, the latter being precluded by the fact that only one proton and one fluorine resonance are observed in the spectrum.

The transient behavior of the protons and fluorines was studied using experiments of type 4 and 5 (see Chapter 6) in which the time dependence of the magnetization at both proton and fluorine resonances was recorded after inverting the magnetization of one spin using a 180° rf pulse. Assuming that the only interactions of importance are those of a proton and fluorine in the same molecule, and noting that protons and fluorines have very similar Larmor frequencies (within 6% of one another), one can set $R_F \approx R_H$ and $S_F(H) \approx S_H(F)$. This assumption is justified by the experimental results shown in Figs. 7-14 and 7-15 where the excellent fit of both proton and fluorine data to the same theoretical curve is shown.

The values of R and S can be directly determined from the transient experiment as discussed in Chapter 6. The determination of the rate constant is

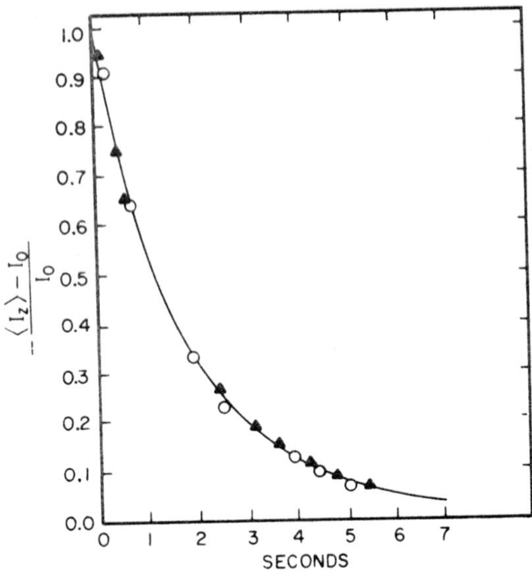

Fig. 7-14   The decay of the magnetization of one kind of nucleus in HF toward its equilibrium value following a 180° rf pulse (Experiment 5 of Chapter 6). O: Fluorine; ▲: Protons;—: Theory. [Solomon (9).]

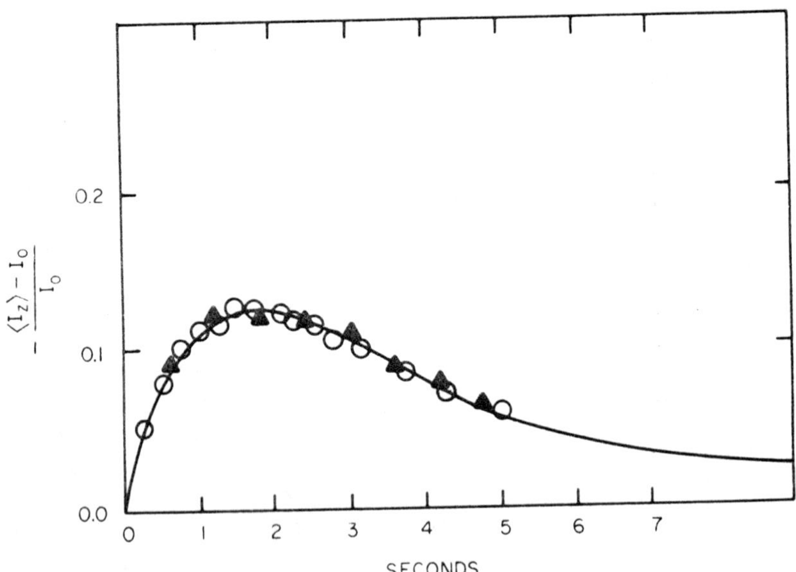

Fig. 7-15   The time dependence of the magnetization of one kind of spin in HF following the application of a 180° rf pulse to the other kind of spin (Experiment 4 of Chapter 6). O: Fluorine; ▲: Protons;—: Theory. [Solomon (9).]

complicated, however, by the fact that the exchange is between magnetically nondistinct sites and only enters the problem through expressions (7.1) and (7.2) for ρ_{scalar} and σ_{scalar}. In addition, $R = \rho_{scalar} + \rho_{dipolar}$ and $S = \sigma_{scalar} + \sigma_{dipolar}$, and the value of J_{HF} was not known because it was not possible to slow the exchange rate to a point where the splitting could be observed in the spectrum. To circumvent these problems, additional experiments were necessary. These included studies of the transverse relaxation time T_2, as well as studies of the longitudinal relaxation time as a function of temperature and of moisture in the sample (since moisture catalyzes proton exchange). The interested reader may refer to the original papers for the details of the analysis. It was determined that the H–F coupling constant was 615 Hz and that the activation energy for exchange of a proton between and HF molecule and an H_3O^+ ion was 1840 ± 250 calmole^{-1}.

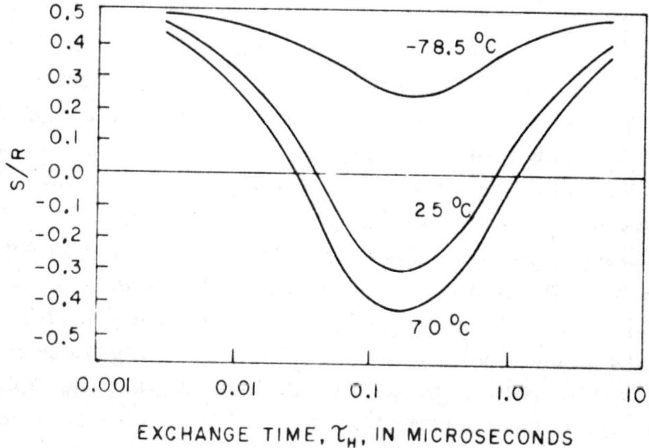

Fig. 7-16 Calculated Overhauser effect as a function of the rate of proton exchange in anhydrous HF at several temperatures. A proton–fluorine scalar coupling constant of 615 Hz was used in the calculation. [Solomon and Bloembergen (10).]

Once the constants determining the behavior of the system had been determined, it was possible to calculate the enhancement as a function of exchange rate as is shown in Fig. 7-16. The values approach $+0.5$ in the limits of slow and fast exchange where the dipole mechanism is dominant. At intermediate rates of exchange, negative enhancements are observed which correspond to a compromise between the $+0.5$ expected from pure dipole relaxation and the -1.0 expected from pure scalar relaxation. The fractional contribution of the scalar mechanism to spin–lattice relaxation g may be obtained from the observed fractional enhancements by noting

$$f = S/R = \frac{\sigma_{\text{dipolar}} + \sigma_{\text{scalar}}}{\rho_{\text{dipolar}} + \rho_{\text{scalar}}}$$

Combined with $\sigma_{\text{scalar}} = -1.0\rho_{\text{scalar}}$, $\sigma_{\text{dipole}} = \tfrac{1}{2}\rho_{\text{dipole}}$, and $g = \rho_{\text{scalar}}/(\rho_{\text{scalar}} + \rho_{\text{dipole}})$, we find

$$g = (1 - 2f)/3 \tag{7.14}$$

The values of f may be read from the graph in Fig. 7-16. At 70°C, for example, we find $0 \leqslant g \leqslant 0.62$. This demonstrates very clearly that the scalar mechanism must be taken into account in interpreting systems with exchanging spins. The reader is referred to the original papers for further discussion of the interpretation of these experiments in terms of the structure and physical chemistry of HF and solutions of HF and its salts.

Another system in which scalar relaxation is important due to the presence of chemical exchange is in alcohols. Grunwald *et al.* (22) first studied the exchange of hydroxyl protons as a function of pH in a methanol system buffered by the addition of a carboxylic acid and its salt. Effects of the buffer components were eliminated by extrapolating to zero buffer concentration. Cocivera (23) then found that the spin–lattice relaxation time in these systems was dependent upon the rate of proton exchange, indicating scalar relaxation, and then proceeded to estimate the rate of exchange under the assumption that the scalar mechanisms was the only significant means of relaxation for hydroxyl proton. This problem was next taken up by Fukumi *et al.* (11) who used the methods of Solomon (9) and Solomon and Bloembergen (10) to obtain estimates of the dipolar contribution as well as the exchange rates. The experiments differed from those done on HF only in that adiabatic rapid passage was used to invert the magnetization instead of the rf pulse method. The problem was also simplified by the fact that J and δ could be obtained directly from the high-resolution spectrum so that temperature studies were not necessary. The solutions were acidified by addition of *p*-nitrobenzoic acid and all measurements were made at 28°C.

Negative Overhauser enhancements were observed between the methyl and hydroxyl proton signals over a wide range of hydrogen ion concentration. This, of course, confirmed the importance of the scalar mechanism. The magnitude of the dipolar relaxation was estimated in a 1.0 M solution of sodium methoxide in methanol where exchange would be much too fast to result in scalar relaxation. The dipole–dipole relaxation measured on this one sample was assumed the same for all samples, allowing the exchange rates to be extracted from the decay constants.

The dipole contribution was estimated to be $\rho_{\text{dip}} = 0.02$ sec^{-1} while the total relaxation rates varied from about 0.1 to 0.28 sec^{-1} for the methyl protons and 0.14 to 0.95 sec^{-1} for the hydroxyl protons, depending upon the

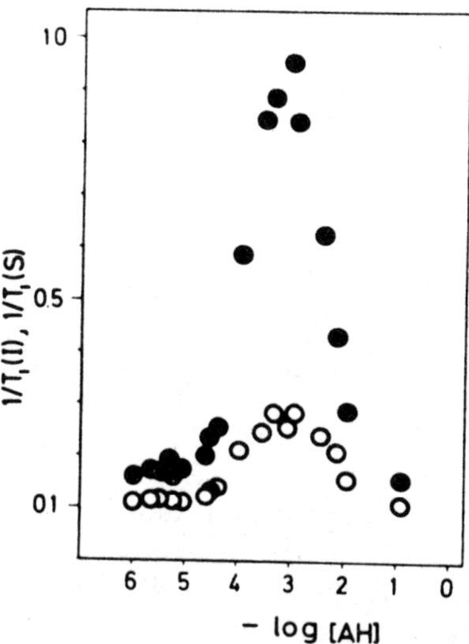

Fig. 7-17 The dependence of the relaxation rate, R (sec^{-1}), of the methyl protons O and the hydroxyl protons ● of methanol on the molarity of acid in a p-nitrobenzoic acid–methanol solution. [Fukumi *et al.* (*11*).]

hydrogen ion concentration. The total relaxation rates as a function of concentration are shown in Fig. 7-17. The fractional enhancements are recorded in Fig. 7-18, from which the relative importance of scalar and dipolar mechanisms may be estimated using Eq. (7.14). The range is $21\% \leqslant g \leqslant 91\%$ for the hydroxyl proton and $21\% \leqslant g \leqslant 73\%$ for the methyl protons.

The predominant reaction is presumed to be

$$CH_3OH_2^+ + CH_3OH \underset{}{\overset{k}{\rightleftharpoons}} CH_3OH + CH_3OH_2^+ \tag{7.15}$$

Using this reaction mechanism, a chemical shift of 152 ± 1 Hz between the

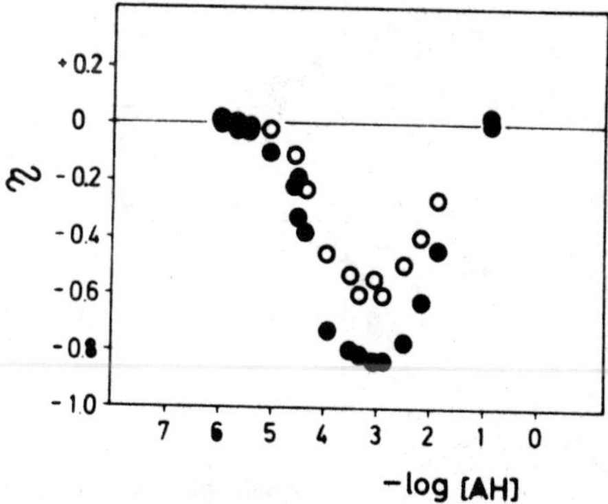

Fig. 7-18 The dependence of the Overhauser enhancements in a solution of p-nitrobenzoic acid in methanol on the acid molarity. ● Enhancement of the methanol –OH proton signal upon saturation of the methanol –CH$_3$ resonance. O Enhancement of the methanol CH$_3$– resonance upon saturating the methanol –OH proton resonance. [Fukumi *et al.* (*11*).]

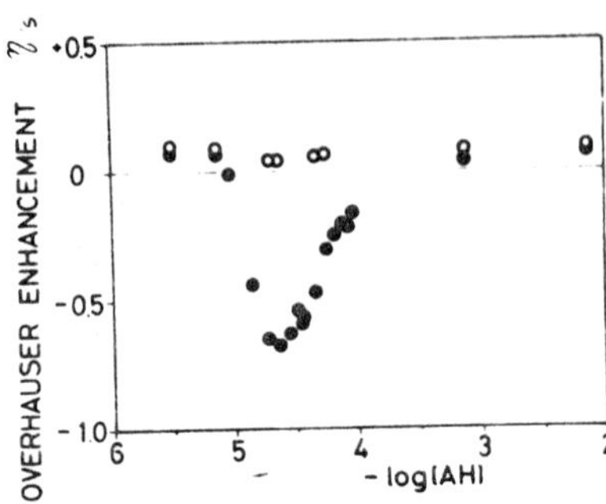

Fig. 7-19   The dependence of the nuclear Overhauser effect on concentration (moles per liter) in solutions of lutidine hydrochloride (AH) in ethanol. ● The enhancement of the ethanol –OH proton resonance upon saturating the ethanol –CH₂– resonance. O The enhancement of the ethanol –OH proton resonance upon saturating the ethanol –CH₃ proton resonance. [Arata *et al.* (*12*).]

methyl and hydroxyl resonances (at 100 MHz), and a coupling constant of 5.1 ± 0.2 Hz, the rate constant k is estimated to be 1.8×10^8 liters mole^{-1} sec^{-1}. Justification for using this mechanism as well as a discussion of the source of the difference between this result and the value $k = 3.5 \times 10^9$ liters mole^{-1} sec^{-1} obtained by Grunwald *et al.* (*22*) are also included in this paper.

Arata *et al.* (*12*) then studied proton exchange in acidified ethanol using the same procedure. Here, however, there are three nonequivalent groups of spins (CH₃–, CH₂–, and –OH) on which the experiments must be done. The fractional enhancements are given as a function of acid concentration in Figs. 7-19 and 7-20. The dependence of the relaxation times on the concentration of lutidine · HCl (added as a buffer) does not fit the model reaction (7.15) used in interpreting the results of the experiments on methanol. A reaction which is consistent with these results is

$$LH^+ + CH_3CH_2OH + L \underset{k}{\rightleftharpoons} L + CH_3CH_2OH + LH^+ \qquad (7.16)$$

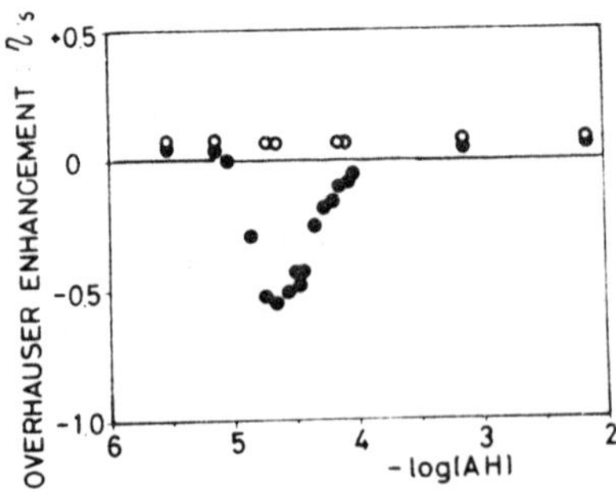

Fig. 7-20   The dependence of the nuclear Overhauser effect on concentration (moles per liter) in solutions of lutidine hydrochloride (AH) in ethanol. ● The enhancement of the –CH₂– proton resonance upon saturating the ethanol –OH proton resonance. O The enhancement of the ethanol –CH₂– resonance upon saturating the ethanol –CH₃ resonance. [Arata *et al.* (*12*).]

where L signifies lutidine. Using this reaction to interpret the kinetic data they find $k = 2 \times 10^9$ liters$^{1/2}$ mole$^{-3/2}$ sec^{-1}.

D. A Review of Applications of NOE to Exchanging Systems: Intramolecular Exchange

One of the earliest applications of double resonance techniques to intramolecular exchange was made by Calder *et al.* (*24*) in conjunction with a study of the temperature dependence of the NMR spectrum of [18] annulene and p-nitrosodimethylaniline (Fig. 7-21). The NMR spectrum of [18] annulene in fully deuterated toluene at $-60°$C consists of a high field quartet ($\tau 14.22$ with respect to tetramethylsilane) due to the six inner protons and a low field quintet ($\tau 0.75$) due to the twelve outer protons. As the temperature is increased

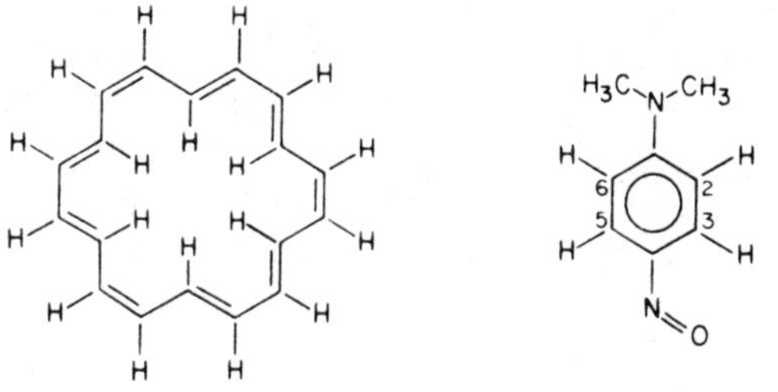

Fig. 7-21   Annulene (A) and *p*-nitrosodimethylaniline (B).

toward $+20°$C, these multiplets broaden and the fine structure is lost. The signals are so broadened as to be undetectable between about $20°$ and $50°$C. At $50°$C a single band is present which narrows as the temperature is increased further. This temperature dependent behavior was assumed to be due to interchange of the inner and outer protons by an internal rotation. This interpretation was confirmed by double-resonance experiments on the sample below $20°$C where two resonances were present. At very low temperatures ($-60°$C), strong irradiation of one multiplet simply resulted in decoupling; the other multiplet collapsed to a singlet but its integral remained unchanged. At $20°$C, however, saturation of one resonance caused a complete loss of signal at the second resonance. Intermediate results were obtained at intermediate temperatures, giving strong confirmation of the mechanism proposed for the temperature dependence of the spectrum. Similar studies have been made on other annulenes (*25, 26*).

The NMR spectrum of *p*-nitrosodimethylaniline in deuterated acetone at $-60°$ consists of a low-field doublet ($\tau 1.12$) due to proton 3, a double doublet ($\tau 3.07$) due to proton 2, and a high-field multiplet ($\tau 3.39$) due to protons 5 and 6. At $-60°C$, decoupling was observed just as with [18] annulene, and the decoupled spectra were consistent with the spectral assignments. At $-20°C$ the results were more interesting. Saturation of the H–5, H–6 band resulted in the disappearance of the low-field H–3 band, while irradiation of the H–3 resonance resulted in disappearance of the H–2 signal as well as the H–5, H–6 signal. Calder *et al.* (*24*) explained these observations in the following manner. Because of the symmetry of the molecule, rotation of the nitroso group converts H–3 into H–5 and H–2 into H–6. At the same time H–5 and H–6 are very strongly coupled so that saturation of H–5 and of H–6 are inseparable. Thus if the H–3 signal is saturated, the exchange of H–3 with H–5 results in saturation of the H–5, H–6 system and, again as a consequence of exchange, saturation of H–5, H–6 leads to saturation of H–2. The direct transfer of saturation between the H–5, H–6 multiplet and H–3 was directly confirmed by the experiment in which H–5, H–6 was saturated. The saturation of H–2 in this manner was not reported, presumably because the proximity of the H–2, and the H–5, H–6 resonances made the experiment impossible.

The interpretation of the results was also confirmed by repeating the experiments in deuterochloroform in which H–5 and H–6 are not tightly coupled. The spectrum at $-60°C$ in this solvent consists of a low field multiplet ($\tau 1.19$) due to H–3, another multiplet ($\tau 3.25$) due to H–2 and H–5 (which are not *J* coupled), and a high field multiplet ($\tau 3.55$) due to H–6. Again, at $-20°C$, irradiation of the H–5, H–2 multiplet results in loss of signal at H–3, but now the saturation of H–3 results in approximately 50% loss of signal from H–5, H–2 and has no effect on the H–6 resonance, as would be expected. In spite of the fact that these systems appear quite amenable to thorough investigation by the double-resonance method, and the fact that data was actually collected over a wide range of temperatures, Calder *et al.* apparently made no attempt to obtain quantitative kinetic data on these systems.

The elegant study of the kinetics of ring inversion in cyclohexane-d_{11} by Anet and Bourn (*27*) clearly shows the quality of results that may be obtained using double-resonance techniques. Prior to the study by Anet and Bourn, the chair–chair interconversion in cyclohexane and cyclohexane-d_{11} had been extensively studied using various NMR techniques (*28–32*). There was general agreement among these studies on the order of magnitude of the rate constant k and thus for the Gibbs's free energy of activation ($\Delta G^{\ddagger}$) for the inversion process. There was some disagreement, however, over the values of the energy of activation ($\Delta E^{\ddagger}$), the enthalpy of activation ($\Delta H^{\ddagger}$), and the entropy of activation ($\Delta S^{\ddagger}$), with the spin–echo results of Allerhand *et al.* (*32*) being notably different than the results of other workers. These activation parameters

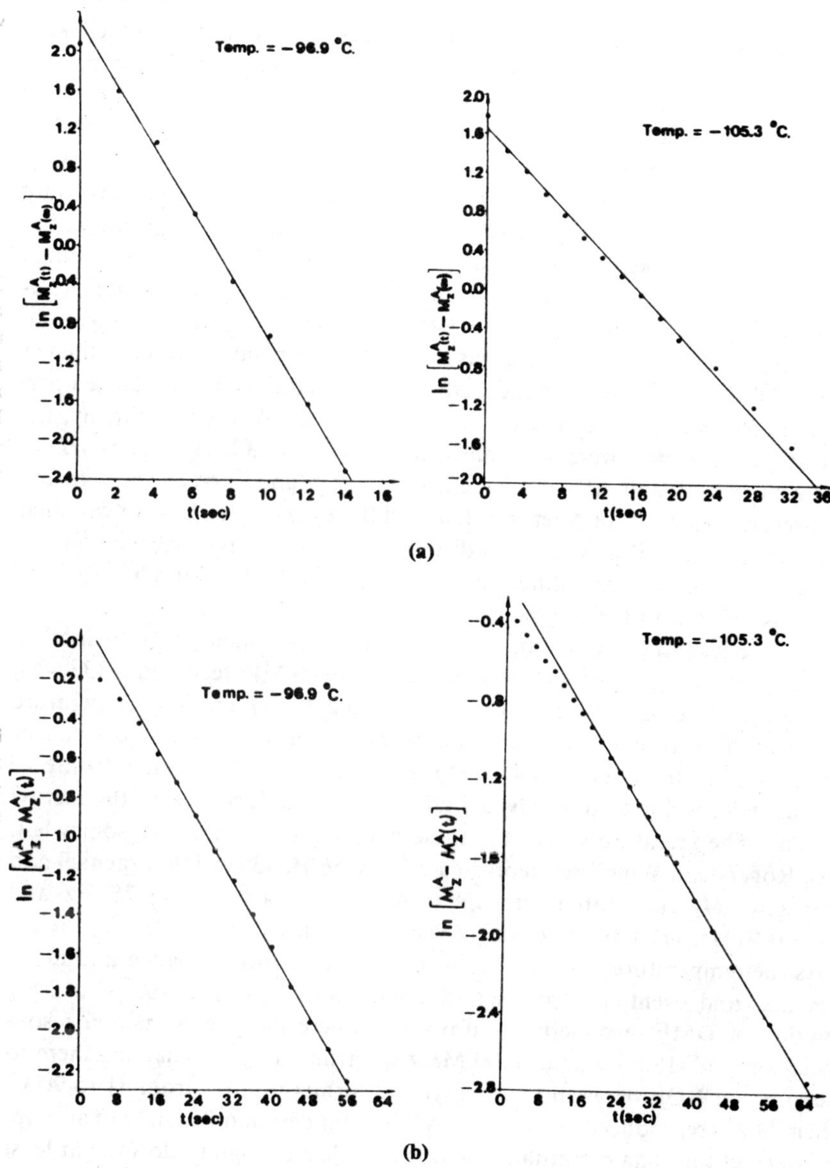

Fig. 7-22 Transient NMR experiments on cyclohexane-d_{11}. (a) Semilogarithmic plots of the decay in the intensity of the high-field proton (A) resonance upon applying a saturating rf field at the frequency of the low-field resonance (B). The slopes of the lines give $R_A = 3.0$ and 9.5 sec at −96.9 and −105.3°C, respectively. (b) Semilogarithmic plots of the recovery of the intensity of the high-field proton resonance after removal of the saturating rf field from the low field resonance. The slopes of the lines give $(R_A + \sigma_A)^{-1} = 23.5$ and 24.8 sec at −96.9 and −105.3°C, respectively. [Reproduced by permission from F. A. L. Anet and A. J. R. Bourn *J. Amer. Chem. Soc.* **89**, 760 (1967).]

are calculated from the temperature dependence of k, so obtaining accurate values for them requires measuring k accurately over as wide a range of temperatures as possible.

Anet and Bourn reinvestigated the kinetics using both lineshape analysis and double-resonance methods. The double-resonance method allowed extension of the study to much lower exchange rates than could be studied by the lineshape technique alone. Experiments of types 1 and 3 were run on the sample over a temperature range of -96.9 to $-116.7°C$, over which range separate resonances are observed for the axial and equatorial protons. Semilog plots of the magnetization versus time for these experiments are shown in Fig. 7-22. This temperature range was found to correspond to rate constants of 0.27 to 0.004 sec^{-1}. The lineshape method was then used to extend the temperature range up to $-24°C$ and the rate constants to 2490 sec^{-1}. Finally, the activation parameters were determined to be $\Delta G^{\ddagger} = 10.3$ kcal $mole^{-1}$, $\Delta H^{\ddagger} = 10.8$ kcal $mole^{-1}$, and $\Delta S^{\ddagger} = +2.8$ eu, which is in good agreement with the high-resolution results of Anet *et al.* (*30*) and Bovey *et al.* (*31*), but not with that of Allerhand *et al.* (*32*). For further discussion of the details of the method and possible reasons for the different results from different laboratories, the reader is referred to the original paper (*27*).

The kinetics of the internal rotation in dimethylformamide (DMF) is another problem which has been extensively studied by NMR techniques (*33–45*), including an Overhauser study by Saunders and Bell (*44*). The low-temperature spectrum of DMF consists of a singlet from the formyl proton and a pair of singlets due to the nonequivalent N–methyl protons. The resonances are of unequal widths due to unresolved J couplings of the formyl with the methyl protons. The literature values for these coupling constants vary somewhat, with Rogers and Woodbrey reporting $J_{BF} = 0.56$ Hz (*35*) while Fraenkel and Franconi (*34*) and Rabinovitz and Pines (*43*) report $J_{BF} = 0.75$ Hz and $J_{AF} = 0.50$ Hz. The labeling of the molecule is defined in Fig. 7-23.

As the temperature is raised the N–methyl resonances broaden and move together, and eventually coalesce. Saunders and Bell (*44*) studied a 2.5% solution of DMF in dimethyl–sulfoxide-d_6 where the two N–methyl resonances were still 15.5 Hz apart (100 MHz spectrum) at 90°C, enabling them to measure the NOE between the methyl and formyl protons from 31 to 90°C. Their data is reproduced in Table 7-4. While Saunders and Bell did not attempt to interpret this data quantitatively, it is complete enough to do so in at least an approximate manner.

The energy of activation $\Delta E^{\ddagger}$ may be estimated from their data under the following assumptions:

(1) The small magnitude of the enhancement of F observed when B is irradiated at low temperatures implies the interaction of the formyl proton with B is much less than its interaction with A. An alternate interpretation

TABLE 7–4

THE NUCLEAR OVERHAUSER EFFECT IN
DIMETHYLFORMAMIDE AS A FUNCTION OF
TEMPERATURE[a]

Temperature (°C)	Enhancement of the resonance of the formyl proton (%)	
	Irradiate $CH_3(A)$	Irradiate $CH_3(B)$
31	28	3
40	—	6
50	28	9
55	—	13
60	28	18
70	28	24
80	28	26
90	28	27–28

[a] Data from Saunders and Bell (44). The enhancements are accurate to $\pm 1\%$ and the temperatures to $\pm 3\%$.

would be that the enhancement of B is canceled by the three spin interaction, $B \leftrightarrow A \leftrightarrow F$ as was discussed in Chapter 3. However, methyl–methyl Overhauser effects (in the absence of direct exchange) are typically very small, and the interpretation to be used here is probably more realistic (cf. p. 168).

(2) The relaxation times for the various groups of spins are essentially independent of temperature over the 30–90°C range studied here. This is supported by the fact that the NOE observed for the formyl proton upon saturating resonance A is independent of temperature. The magnitude of this NOE would also be constant, however, if the interaction with A were the only source of relaxation for the formyl proton, but in that case the enhancement would be expected to be 50%. Note that the assumption that the interaction with B is negligible combined with the fact that the observed enhancement when A is saturated is only 28% requires that intermolecular relaxation of the formyl proton be approximately as efficient as its interaction with A.

As a consequence of assumption (1), the NOE observed for F when B is saturated must be a secondary effect. The primary effect of saturating resonance B is to cause partial saturation of resonance A due to exchange of these spins. The saturation of A results in turn in an enhancement of the F signal. This may be described quantitatively as

$$f_F(A) = \frac{\langle \mathbf{I}_{zF} \rangle - I_{0F}}{I_{0F}} = -\frac{\sigma_{FA}}{R_F} \frac{\langle \mathbf{I}_{zA} \rangle - I_{0A}}{I_{0A}} \tag{7.17}$$

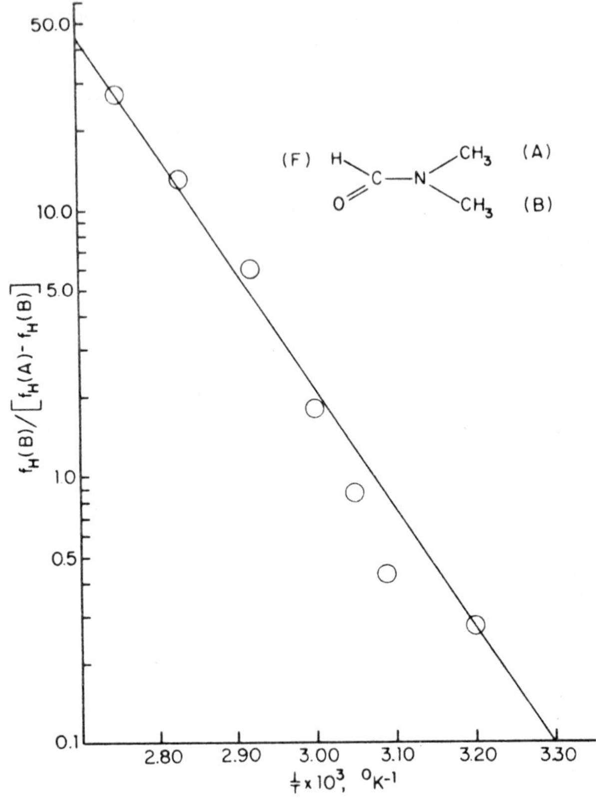

Fig. 7-23 A representation of the temperature dependence of the nuclear Overhauser effect in dimethylformamide. The measurements were made on a 2.5% solution of dimethylformamide in dimethylsulfoxide-d_6. The slope of the line indicates an activation energy of about 8.7 kcal mole^{-1} for rotation about the C–N bond. [NOE data from J. K. Saunders and R. A. Bell *Can. J. Chem.* **48**, 512 (1970). Reproduced by permission of the National Research Council of Canada.]

Using the results of Section B of this Chapter and assuming B is completely saturated

$$f_A(B) = \frac{\langle \mathbf{I}_{zA} \rangle - I_{0A}}{I_{0A}} = \frac{-k}{2W_{1A} + k} \tag{7.18}$$

where k is the rate constant for exchange of A and B. Combining these gives

$$f_F(B) = \frac{\sigma_{FA}}{R_F} \frac{k}{2W_{1A} + k} \tag{7.19}$$

But from (7.19) and (7.17) we also see that, if A were completely saturated, $f_F(A) = \sigma_{FA}/R_F$, so we may rewrite (7.19) as†

$$f_F(B) = f_F(A)\frac{k}{2W_{1A} + k} \tag{7.20}$$

If we now assume $k = k_0 \exp(-\Delta E^{\ddagger}/RT)$, where R is the gas constant, substituting into (7.20) and rearranging yields

$$\frac{f_F(B)}{f_F(A) - f_F(B)} = \frac{k_0}{2W_{1A}} \exp\left(\frac{-\Delta E^{\ddagger}}{RT}\right) \tag{7.21}$$

Assumption 2 allows us to treat W_{1A} as a constant, so a semilog plot of $f_F(B)/(f_F(A)-f_F(B))$ should be a straight line with a slope of $-\Delta E^{\ddagger}$. This plot is shown in Fig. 7-23. Estimating the slope from this plot we obtain $\Delta E^{\ddagger} \approx 8.7$ kcal/mole. The preexponential obtained in this way is $k_0/2W_{1A} \approx 10^{13.4}$. Assuming $2W_{1A} \approx 10^{-1}$, this would imply $k_0 = 10^{14}$. Values of $\Delta E^{\ddagger}$ and k_0 selected from the literature are summarized in Table 7-5 for comparison.

TABLE 7-5

SUMMARY OF THE ACTIVATION ENERGIES AND PREEXPONENTIAL FACTORS
REPORTED FOR THE INTERNAL ROTATION IN DIMETHYLFORMAMIDE

Solvent	$\Delta E^{\ddagger}$ (kcal/mole)	$\log k_0$	Reference
Neat	7. $\pm$3.	3 to 7	33
Neat	9.6$\pm$1.5	6.5	34
Neat	18.3$+$0.7[a]	10.8$\pm$0.4	35
Neat	15.9$\pm$2.0	—	37
Neat	20.5$\pm$	12.7	43
0.4 M in 100% H_2SO_4	12.7$\pm$1.5	8.0	34
2.5% in DMSO-d_6	8.7	14	b
0.0858 mole % in acetone-d_6	16.8$\pm$2.0	—	37
0.105 mole % in $CFCl_3$	11.3$\pm$2.0	—	37
0.0633 mole % in HMDS[c]	9.4$\pm$1.0	—	37
0.6 M in a 1:1 mixture of 1-chloronapthalene and benzotrichloride	23.2$\pm$2.0	13.3$\pm$1.1	41

[a] This value would be adjusted upward if appropriate corrections were made for the spin–spin splitting and overlap of the methyl resonances.
[b] Estimated from the data of Saunders and Bell (44).
[c] Hexamethyldisiloxane.

---

† This effect is quite analogous to the three-spin effect discussed in Section 3, G [Eq. (3.39)] except that, in this case, one of the couplings is by exchange rather than by dipole–dipole relaxation.

TABLE 7-6

RATE CONSTANTS FOR THE INTERNAL ROTATION IN
DIMETHYLFORMAMIDE[a]

Temperature (°C)	k (sec^{-1})	Temperature (°C)	k (sec^{-1})
80	0.94	107	6.91
85	1.21	111	9.61
90	2.32	114.5	11.99
94	2.70	119[b]	15.61
99	3.89	122	20.35
103	5.23	128	29.20

[a] Data taken from Rabinovitz and Pines (*43*).
[b] Coalescence temperature of the N-methyl resonances.

It can be seen that our estimate is not out of line with other reported values
for these parameters. It is rather surprising to see as great a scatter in reported
results as Table 7-5 shows. There are several factors that may contribute to this
scatter. The first is that proper corrections have not always been made for the
effect of the scalar couplings, J_{AF} and J_{BF}, on the linewidths and shapes.
The range of rates covered in these studies (see Table 7-6) also fall in a region
where scalar relaxation might be significant, particularly if the relaxation times
are quite long. Another problem which is ignored in most of these studies is
that DMF exists partly as dimers with the structure shown in Fig. 7-24
(*42, 46, 47*). Finally, it is quite possible that the exchange process is catalyzed
or at least strongly affected by traces of moisture or other uncontrolled
impurities. A very careful study of DMF will probably be required to resolve
the question of the kinetics of its internal rotation and the reason for the
differences between the results of different workers. It is a good example of the
type of problem where the combined use of NOE and lineshape methods to
obtain data on a single sample over a wide range of rates would be most
useful, particularly if combined with close attention to sample purity.

A final example of the application of Overhauser experiments to problems
in intramolecular exchange can be taken from the work of Van Leeuwen and

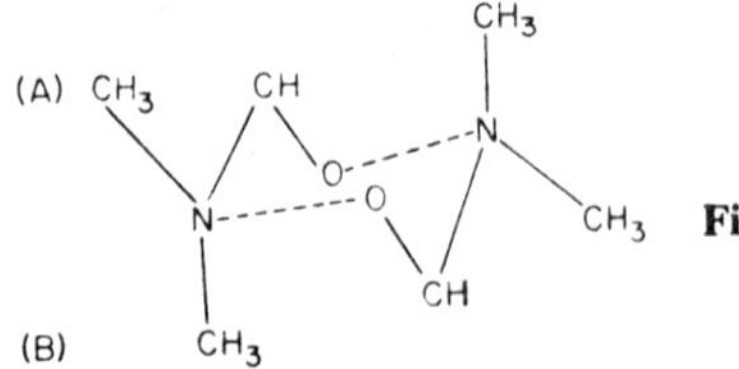

Fig. 7-24   Structure of the dimethylformamide dimer.

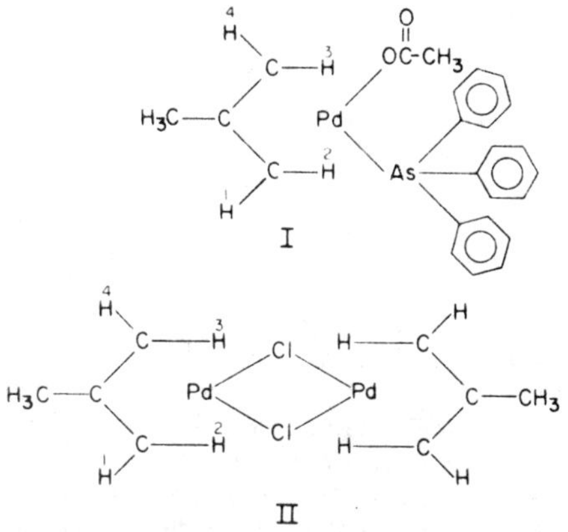

Fig. 7-25 Structure of two π-allylpalladium complexes.

Praat on π-allylpalladium complexes (48). The complexes they reported on are shown in Fig. 7-25. In compound I [and analogs (49, 50)] protons 1 through 4 are nonequivalent and show distinct NMR resonances. Saturating one resonance while observing the others showed that proton 1 exchanges with proton 4, and proton 2 exchanges with proton 3. In a solution of I in $CDCl_3$ these workers estimated a value of 1 sec^{-1} for the interchange rate constant, presumably at $-30°C$. This value will be of very little use, however, because the interchange of protons is believed to result from exchange of ligand between the free and complexed state, but no data were given on the concentration of free ligand or complex in their sample.

In compound II exchange occurs between the *syn* and *anti* protons, but is too slow to be detected by any NMR method other than the NOE method. Using experiments of types 1 and 3, the order of magnitude of the interchange rate was estimated to be 0.2 sec^{-1} at 90°C, with proton relaxation times on the order of 10 sec. Van Leeuwen and Praat's report is very sketchy, however, and little confidence can be placed in their results. It does point out that the full power of NMR methods for studying relatively slow exchange processes has not been exploited by the inorganic chemist to the extent that might be expected.

REFERENCES

1. A general review of exchange effects in NMR may be found in C. S. Johnson, Jr.. *Advan. Mag. Resonance* **1**, 33 (1965).
2. C. P. Slichter, "Principles of Magnetic Resonance," p. 149. Harper, New York, 1963.
3. A. Abragam, "The Principles of Nuclear Magnetism," p. 315. Oxford Univ. Press, London and New York, 1961.
4. P. L. Corio, "Structure of High Resolution Nuclear Magnetic Resonance Spectra," Chapter 5, Sect. D. Academic Press, New York, 1966.

5. A. G. Marshall, *J. Chem. Phys.* **52**, 2527 (1970).
6. S. Forsén and R. A. Hoffman, *J. Chem. Phys.* **40**, 1189 (1964).
7. S. Forsén and R. A. Hoffman, *Acta Chem. Scand.* **17**, 1787 (1963).
8. S. Forsén and R. A. Hoffman, *J. Chem. Phys.* **39**, 2892 (1963).
9. I. Solomon, *Phys. Rev.* **99**, 559 (1955).
10. I. Solomon and N. Bloembergen, *J. Chem. Phys.* **25**, 261 (1956).
11. T. Fukumi, Y. Arata, and S. Fujiwara, *J. Chem. Phys.* **49**, 4198 (1968).
12. Y. Arata, T. Fukumi, and S. Fujiwara, *J. Chem. Phys.* **51**, 859 (1969).
13. A. A. Westenberg and N. de Haas, *J. Chem. Phys.* **48**, 4405 (1968).
14. J. H. Sullivan, *J. Chem. Phys.* **51**, 2288 (1969).
15. J. Feeney and A. Heinrich, *Chem. Commun.* p. 295 (1966).
16. B. M. Fung and R. D. Stolow, *Chem. Commun.* p. 257 (1967).
17. R. A. Hoffman and S. Forsén, *Progr. NMR Spectros.* **1**, 15 (1966).
18. B. M. Fung, *J. Chem. Phys.* **47**, 1409 (1967).
19. B. M. Fung, *J. Amer. Chem. Soc.* **90**, 219 (1968).
20. B. M. Fung, *J. Chem. Phys.* **49**, 2973 (1968).
21. L. W. Reeves and W. G. Schneider, *Can. J. Chem.* **36**, 793 (1958).
22. E. Grunwald, C. F. Jumper, and S. Meiboom, *J. Amer. Chem. Soc.* **84**, 4664 (1962).
23. M. Cocivera, *J. Chem. Phys.* **47**, 1112 (1967).
24. I. C. Calder, P. J. Garratt, and F. Sondheimer, *Chem. Commun.* p. 41 (1967).
25. I. C. Calder, P. J. Garratt, H. C. Longuet-Higgins, F. Sondheimer, and R. Wolovsky, *J. Chem. Soc. C* p. 1044 (1967).
26. I. C. Calder, Y. Gaonie, P. J. Garratt, and F. Sondheimer, *J. Amer. Chem. Soc.* **90**. 4954 (1968).
27. F. A. L. Anet and A. J. R. Bourn, *J. Amer. Chem. Soc.* **89**, 760 (1967).
28. F. R. Jensen, D. S. Noyce, C. H. Sederholm, and A. J. Berlin, *J. Amer. Chem. Soc.* **82**, 1256 (1960), **84**, 836 (1962).
29. R. K. Harris and N. Sheppard, *Proc. Chem. Soc.* p. 419 (1961).
30. F. A. L. Anet, M. Ahmed, and L. D. Hall, *Proc. Chem. Soc. London* p. 145 (1964).
31. F. A. Bovey, F. P. Hood, E. W. Anderson, and R. L. Kornegay, *Proc. Chem. Soc. London* p. 146 (1964), *J. Chem. Phys.* **41**, 2041 (1964).
32. A. Allerhand, F. Chen, and H. S. Gutowsky, *J. Chem. Phys.* **42**, 3040 (1965).
33. H. S. Gutowsky and C. H. Holm, *J. Chem. Phys.* **25**, 1228 (1956).
34. G. Fraenkel and C. Franconi, *J. Amer. Chem. Soc.* **82**, 4478 (1960).
35. M. T. Rogers and J. C. Woodbrey, *J. Phys. Chem.* **66**, 540 (1962).
36. F. A. L. Anet and A. J. R. Bourn, *J. Amer. Chem. Soc.* **87**, 5250 (1965).
37. A. G. Whittaker and S. Siegel, *J. Chem. Phys.* **42**, 3320 (1965).
38. C. W. Fryer, F. Conti, and C. Franconi, *Ric. Sci. Rend.* **A8**, 788 (1965).
39. F. Conti and W. von Phillipsborn, *Helv. Chim. Acta* **50**, 603 (1967).
40. A. Mannschreck, *Tetrahedron Lett.* p. 1341 (1965).
41. A. Mannschreck, A. Mattheus, and G. Rissman, *J. Mol. Spectros.* **23**, 15 (1967).
42. A. Pines and M. Rabinovitz, *Tetrahedron Lett.* p. 3529 (1968).
43. M. Rabinovitz and A. Pines, *J. Amer. Chem. Soc.* **91**, 1585 (1969).
44. J. K. Saunders and R. A. Bell, *Can. J. Chem.* **48**, 512 (1970).
45. J. V. Hattone and R. E. Richards, *Mol. Phys.* **5**, 139 (1962).
46. J. C. Woodbrey and M. T. Rogers, *J. Amer. Chem. Soc.* **84**, 13 (1962).
47. R. C. Neuman, Jr. and L. B. Young, *J. Phys. Chem.* **69**, 2570 (1965).
48. P. W. N. M. Van Leeuwen and A. P. Praat, *J. Organometal. Chem.* **22**, 483 (1970).
49. P. W. N. M. Van Leeuwen and A. P. Praat, *Chem. Commun.* p. 365 (1970).
50. P. W. N. M. Van Leeuwen and A. P. Praat, *J. Organometal. Chem.* **21**, 501 (1970).

APPLICATIONS OF THE NUCLEAR OVERHAUSER EFFECT: A REVIEW OF THE LITERATURE

The number of applications of the Overhauser effect reported in the literature has increased rapidly during the last few years. The NOE studies of chemical kinetics that have appeared have already been reviewed in Chapter 7. In this chapter we shall review the applications that have been made to spectral assignment and to the elucidation of molecular structures and conformations. (A complete set of structural formulas for the compounds discussed in this chapter can be found beginning on page 208.) The reader will find that most of these applications have been to structural problems in natural product chemistry. The many potentially fruitful applications in other areas of chemistry (e.g., conformational problems in organometallic chemistry) have not been investigated at all. We might expect to see rapid growth in these areas in the near future, particularly with the advent of commercial instruments capable of heteronuclear NOE experiments.

A. Assignment of Lines in NMR Spectra

One of the first uses of the nuclear Overhauser effect to be reported in the literature was the assignment of lines in an NMR spectrum. It is usually possible to assign most of the lines in the spectrum of a compound of known structure on the basis of their chemical shifts, coupling constants, integrals, and the results of spin–decoupling experiments. However, there may be several lines in the spectrum which are known to arise from spins with very different positions in the molecule but whose assignments cannot be made on the basis of such information. As was first pointed out by Anet and Bourn (I), it may be possible to obtain an assignment of the spectrum in these cases by taking advantage of the unique spatial dependence of the NOE. Anet and Bourn demonstrated the method on compounds **1–3**, for which spectral assignments had been made by other means.

β,β-Dimethylacrylic acid (**1**)

Compound **1** was studied as a 9% solution in benzene-d_6 with 9% benzene added as an internal reference for field-frequency locking. The spectrum consisted of three multiplets: a 1:6:15:20:15:6:1 septet at $\tau = 4.34$ due to H–1; a doublet with $J = 1.3$ Hz at $\tau = 8.03$ due to one of the methyls; and another doublet with $J = 1.3$ Hz at $\tau = 8.58$, due to the other methyl. The question was which methyl corresponded to the upfield doublet and which to the downfield doublet. In order to answer this question, NOE experiments were performed in which one or the other of the methyl doublets was saturated while H–1 was observed.

Irradiation of either of the methyl doublets resulted in the reduction of the H–1 septet to a 1:3:3:1 quartet. However, when the high-field methyl ($\tau = 8.58$) was irradiated, the integral of the H–1 resonance increased by $17 \pm 1\%$ whereas saturation of the low field ($\tau = 8.03$) methyl resulted in a $4 \pm 1\%$ reduction in the intensity of the H–1 resonance. As the A methyl group is much closer to H–1 than the B methyl group, the large enhancement of the H–1 signal when the upfield doublet was saturated indicated that the upfield resonance was that of the A methyl protons. This was in agreement with assignments made earlier on the basis of chemical shifts (2). The small decrease in the intensity of H–1 when the B methyl was irradiated was probably the first reported case of a negative enhancement arising from a three spin interaction. The magnitude of the effect (4%) in this case was, however, too small to rule out the possibility of its being an instrumental effect.

Dimethylformamide (**2**)

In dimethylformamide one again faces the problem of assigning two methyl peaks which are not clearly identified on the basis of chemical shifts or coupling constants. An NOE study on an 8% solution of **2** in deuterium oxide provided unequivocal assignment of the spectrum. Saturation of the low field methyl produced an $18 \pm 1\%$ enhancement of the signal of the formyl proton, while saturation of the high-field methyl resulted in a $2 \pm 1\%$ decrease in the signal intensity. It was thus concluded, in agreement with previous assignments (*3–5*), that the low field resonance was that of the methyl cis to the formyl proton. Again, a small negative three spin interaction was observed (also, cf. p. 161).

Half-Cage Acetate (3)

The last molecule discussed by Anet and Bourn in their communication was the "half-cage acetate" (3). A 45% enhancement of the signal from H–B when H–A was saturated had been reported in a separate communication (*6*). It was noted that the intensity of either line increased when the other was saturated, even in samples which had not been purged of dissolved oxygen. These observations show that the mutual dipole–dipole interaction between protons A and B is much stronger than their interaction with any other spins present, including the electrons of paramagnetic oxygen in the solution. For this to be the case, A and B must be very close together indeed.

Citral a (4) and Citral b (5)

Ohtsuru *et al.* (*7*) used the NOE to assign the spectra of citral a and citral b as part of a study in which they confirmed the configurations and conformations previously assigned to these compounds. The NMR spectra and their assignments are shown in Figs. 8-1 and 8-2. The saturation of CH_3–f in citral a gave no observable enhancement of the H–g resonance, while in citral b the resonance of H–g was enhanced 18%. As the largest enhancement would be expected from the methyl group closest to H–g, these results confirmed the fact that CH_3–f was cis to the –CHO– group in citral a and trans to it in citral b. The assignment of the methyl resonance at $\tau = 8.32$ to CH_3–b was made by noting that in both compounds saturation of the line led to a 16% enhancement of the signal from H–c. Ohtsuru *et al.* did not report the effect on H–c of saturating the methyl resonance at $\tau = 8.39$. While the 16% enhancement observed does make their assignment likely, it cannot be considered conclusive. CH_3–b would be expected to give a larger enhancement of the H–c resonance than would CH_3–a, but this means that the results of both experiments must be known in order to make the assignment. The importance of using comparative data in arriving at a conclusion rather than using only the absolute magnitude of a single NOE cannot be overemphasized.

1,2,3,4-Tetramethylphenanthrene (6)

Martin and Nouls (*8*) studied solutions containing 12.5% of **6** in deuterated chloroform with TMS added for field-frequency locking. The important features of the 60-MHz spectrum were the H–10 doublet at 7.94 ppm, the H–5 multiplet at 8.52 ppm, and the methyl resonances at 2.43, 2.63, and 2.88 ppm.

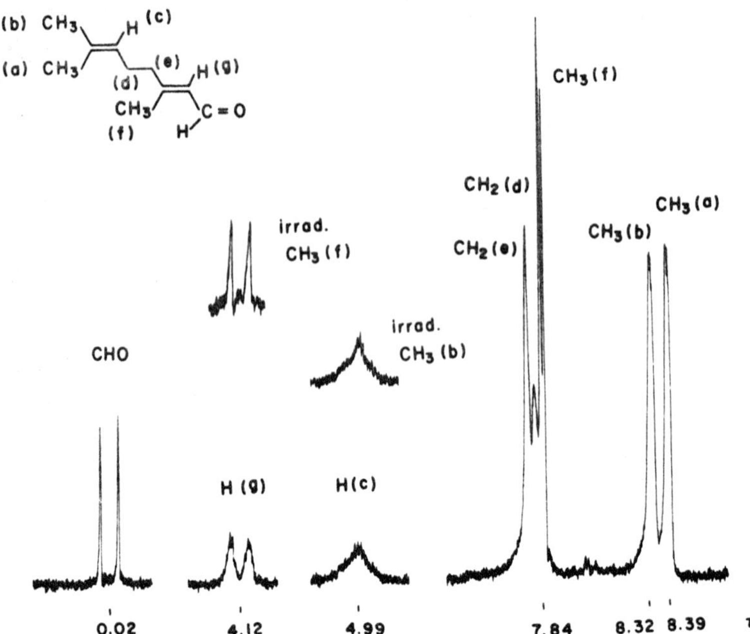

Fig. 8-1 The 100-MHz NMR spectrum of citral a in CDCl₃. [Ohtsuru *et al.* (7).]

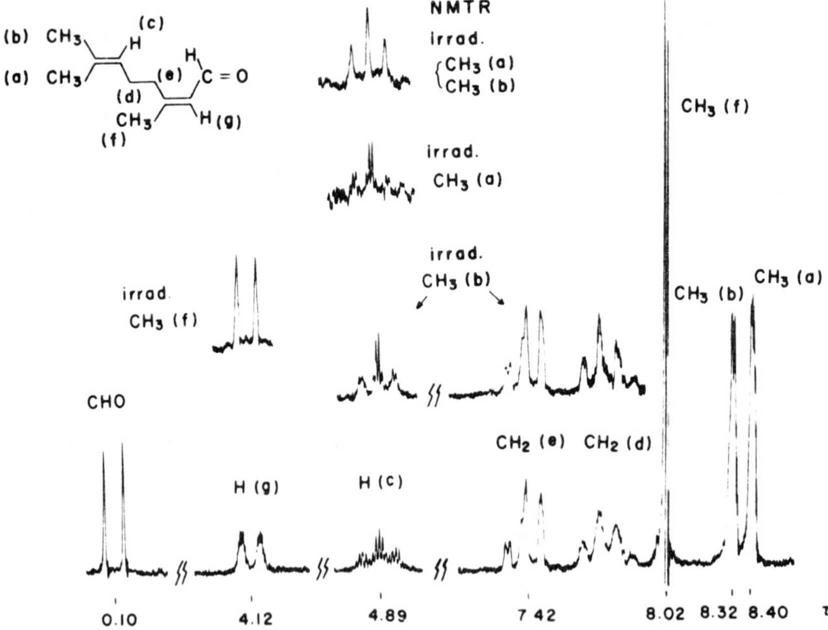

Fig. 8-2 The 100-MHz NMR spectrum of citral b in CDCl₃. [Ohtsuru *et al.* (7).]

The methyl resonance at 2.43 ppm had an integral equivalent to six protons and was presumably due to accidental coincidence of the resonances of the 2- and 3-methyl groups, and this was confirmed by the fact that saturation of this line had no effect on the intensities of the H–5 or H–10 resonances. On the other hand, the saturation of the methyl line at 2.63 ppm resulted in an $11 \pm 3\%$ increase in the intensity of H–10, while saturation of the methyl at 2.88 ppm resulted in a $33 \pm 3\%$ enhancement of the H–5 resonance. The resonance at 2.63 ppm is thus assigned to CH_3–1 and that at 2.88 ppm to CH_3–4.

Camphene (7)

Grover and Stothers (9) studied a 5 mole % solution of camphene in carbon tetrachloride at 60 MHz. The low-field methylene proton signals at 4.65 and 4.44 ppm were assigned to the *syn-* and *anti-exo*protons, respectively, by use of the NOE. Saturating the methyl signals at 1.01 and 1.05 ppm led to a 16% enhancement of the band at 4.44 ppm but no change in the band at 4.65 ppm, thus providing an unequivocal assignment of these resonances.

Caffeine (8)

As part of their study of the self-association of caffeine (8) in aqueous solution, Thakkar *et al.* (10) used the NOE to partially confirm the assignment of the spectrum. The NMR spectrum of a 0.01 M solution of caffeine in D_2O contains four signals of relative intensity 1:3:3:3 at about 8.31, 4.36, 3.93, and 3.75 ppm from TMS. These signals had previously been assigned to H–8, and the 7-, 3-, and 1-methyl protons, respectively (*11–13*). The assignment of the 7-methyl was confirmed by observing an enhancement of the H–8 signal when the line at 4.36 ppm was saturated. No results were reported for experiments in which the methyl lines at 3.93 ppm were saturated while H–8 was observed, nor was the magnitude of the enhancement of H–8 by the 7-methyl protons reported.

Theaflavin (9)

Theaflavin (9) is an orange-red crystalline material which can be extracted from black tea. The determination of its structure relied heavily upon the use of NMR and mass spectroscopy (*14, 15*). Although it made no difference in the structure determination, Bryce *et al.* (16) claimed that there was an error in the original assignment of the spectrum. The protons on the benzotropolone residue give rise to three singlets located (in deutero-chloroform) at 7.55, 7.95, and 8.05 ppm, presumably with respect to TMS. These lines had originally been assigned to protons *C*, *B*, and *A*, respectively. However, Bryce *et al.* observed an NOE (the magnitude of the NOE was not reported) between the line at 7.97 ppm and the H–2' resonance, and, as inspection of models showed that *B* approached H–2' closer than either of the other two benzotropolone aromatic protons, they reassigned the spectrum so that the lines at 7.55, 7.95, and 8.05 ppm corresponded to spins A, B, and C, respectively. This conclusion is not warranted, however, in the absence of NOE experiments between the line at 8.05 ppm and the H–2' resonance. If the lines at 7.55 and 8.05 ppm were each saturated while observing H–2', the line which gave the largest enhancement would correspond to the spin closest to H–2'. Then an examination of models could establish the assignment with a high level of confidence. The assignment given by Bryce *et al.* was, nonetheless, assured of being correct because they also found long-range benzylic couplings between H–2 and H–A, and between H–2' and H–B, which previous workers had not reported.

Assignment of ^{13}C Spectra by Heteronuclear NOE

Jones *et al.* (*17*) have reported a procedure for using ^{1}H–^{13}C NOE's to assign lines in the ^{13}C NMR spectrum arising from nonequivalent quaternary carbon atoms. The experiment consists of observing the enhancements of the unassigned lines in the ^{13}C spectrum while saturating all proton resonances using a white noise decoupler. The observed enhancements are interpreted by comparing them with a set of calculated enhancements. It is very easy to calculate the enhancements needed in this procedure. The equation used is

$$f_j(\text{H}) = \tfrac{1}{2}(\gamma_\text{H}/\gamma_\text{C})/[\alpha/(\alpha + \chi)] \tag{8.1}$$

where $f_j(\text{H})$ is the enhancement of the resonance of carbon j when all proton lines are saturated,

$$\alpha = \sum_\text{H} r_{j\text{H}}^{-6}, \qquad \chi = \rho_\text{C}^*/\gamma_\text{H}^2 \gamma_\text{C}^2 \hbar^2 \tau_\text{C},$$

and the sum runs over all protons in the molecule. Equation (8.1) follows directly from Eq. (3.6) under the following assumptions: (1) all proton lines are completely saturated, (2) the correlation time for all interactions is the same, and (3) no significant fraction of the ^{13}C nuclei have other ^{13}C nuclei as near neighbors in the molecule. The third assumption allows the second term in Eq. (3.6) to be dropped. As the intermolecular contribution to the spin relaxation is not generally known, Jones *et al.* proceed by plotting $f_j(\text{H})$ versus χ for the carbons of interest and then looking for the value of χ which gives the best fit to the experimental data. This procedure may be illustrated using fluoranthene as an example.

Fluoranthene (10)

The calculated values of $f_j(\text{H})$ versus χ for several of the carbons in fluoranthene are shown in Fig. 8-3. In performing the calculations, Jones *et al.* (*17*) assumed that χ was the same for all carbon nuclei in the molecule. The values

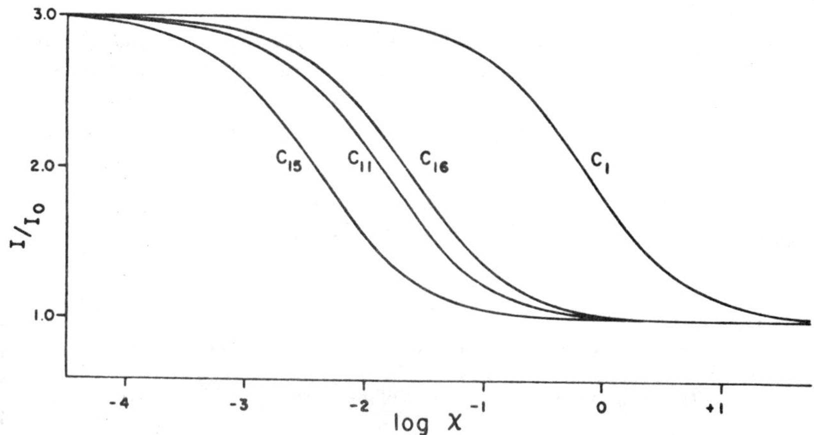

Fig. 8-3 The $^{13}\text{C}-\{^1\text{H}\}$ Overhauser effect in fluoranthene (10) as a function of the intermolecular spin–lattice relaxation rate. The enhancements are expressed as $I_j/I_{0j} = 1 + f_j(\text{H})$, and $\chi = \rho^*/\gamma_H^2 \, \gamma_C^2 \hbar^2 \tau_C$. [Reproduced by permission from A. J. Jones, D. M. Grant, and K. F. Kuhlmann, *J. Amer. Chem. Soc.* **85**, 1010 (1963).]

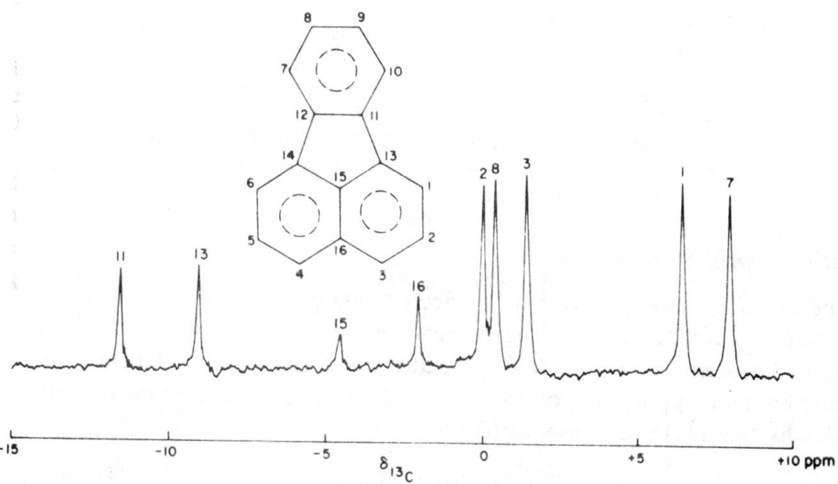

Fig. 8-4 The proton decoupled ^{13}C NMR spectrum of fluoranthene (10). [Reproduced by permission from A. J. Jones, D. M. Grant, and K. F. Kuhlmann, *J. Amer. Chem. Soc.* **85**, 1010 (1963).]

of $f_j(\text{H})$ range from 1.98, corresponding to the maximum possible enhancement (see Table 3-8), down to 0.0, or no enhancement at all. The significant point is that in all cases where an enhancement would be observed, the magnitudes of the enhancements fall in the order

$$f_{C1}(\text{H}) \geqslant f_{C16}(\text{H}) \geqslant f_{C11}(\text{H}) \geqslant f_{C15}(\text{H})$$

Thus the relative magnitudes of the NOE's allows unequivocal assignment of the lines in the spectrum. The assigned ^{13}C spectrum of fluoranthene is shown in Fig. 8-4, and the NOE data is presented in Table 8-1. The assignments of the ^{13}C spectra of acenapthylene (**11**), acenapthene (**12**), and pyrene (**13**) were made in the same manner (*17*).

This method has also been used to distinguish between C–9 and C–10 in 1,8-dimethylnapthalene (**14**) in a paper which also reviews much of the data just discussed (*18*).

TABLE 8-1

^{13}C–$\{^{1}H\}$ OVERHAUSER EFFECTS IN FLUORANTHENE (**10**)[a]

Resonance Observed	Percentage Enhancement[b]
C–1 and C–6	1.82
C–2 and C–5	1.82
C–3 and C–4	1.82
C–7 and C–10	1.82
C–8 and C–9	1.82
C–11 and C–12	1.35
C–13 and C–14	1.35
C–15	1.12
C–16	1.52

[a] Data from Jones *et al.* (*17*).

[b] The enhancements of the ^{13}C resonances were observed while saturating all of the proton resonances using a "white noise" decoupler.

Tightly Coupled Systems

The use of generalized Overhauser effects to assign spectral lines to specific spin transitions or for determining the relative signs of coupling constants are beyond the scope of this book. The reader is referred to the literature for discussion and application of these procedures [e.g., see Kuhlmann and Baldeschwieler (*19*) and Kaiser (*20*)].

B. Determination of Molecular Configuration

The magnitude of the Overhauser effects observed in a molecule is determined by the relative internuclear distances and thus may be used to determine which of several spins are closest to another spin. If the structure of the molecule is known, then determining which lines in its NMR spectrum arise from nuclei which are close together may allow the spectrum to be assigned, as was discussed in Section A. On the other hand, if the assignment of the spectrum is known, then determining which nuclei are close together by using the NOE may establish the configuration of the molecule. The cases where configuration assignments have been made in this manner are the subject of the review in this section.

A good example for illustrating this technique is found in the work of Nouls *et al.* (*21*) on assigning geometries to the two isomeric 3-ethylidene-1-azabicyclo [2,2,2] octanes (**15, 16**) which are obtained by the Wittig reaction of triphenylphosphorane on 3-quinuclidone (*22*). Figure 8-5 shows the 100 MHz spectrum of a mixture containing about 70% isomer *a* and 30% isomer *b*

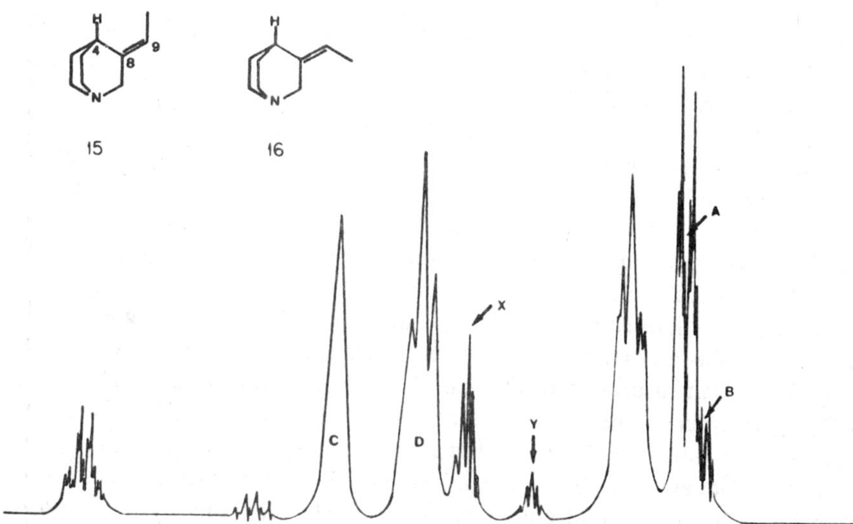

Fig. 8-5 The 100-MHz proton NMR spectrum of a 70:30 mixture of 3-ethylidene-1-azabicyclo (2,2,2) octanes, **15** and **16**. [Nouls *et al.* (*21*).]

in $CDCl_3$ with 10% by weight CF_3COOH added for field-frequency locking. The proportions of the two isomers were determined from the relative signal amplitudes. The H–4 resonance is labeled X in isomer *a* and Y in isomer *b*. The 9-methyl resonance is labeled A in *a* and B in *b*. It can be seen in Fig. 8-5 that this region of the spectrum is very similar for both molecules, but the spectrum of *a* is shifted from that of *b* so that the lines A, B, X, and Y are all well isolated from one another and NOE experiments can be done between them. Two experiments were reported: saturating B had no effect on the intensity of Y, while saturating A resulted in a 31% increase in the intensity of X. Nouls *et al.* were able to conclude from this that H–4 and the 9-methyl are closest together in isomer *a*, so that *a* corresponds to the 9-methyl, H–4 cis isomer, compound **15**. This was in agreement with their earlier conclusion that the major product of the Wittig reaction of triphenylethylphosphorane on 3-quinuclidone was the cis isomer.

Colson *et al.* (*23*) applied the NOE in a similar manner to make stereochemical assignments for several 7,12-dihydropleiadenes. They first tested the procedure on compounds **17–19** for which the stereochemistry was already known. The experiments were performed at 100 MHz on solutions in CDCl$_3$. In compound **17**, ring inversion is impossible, so saturation of the 12-methyl group would be expected to give a rather large enhancement of the signal due to the axial H–7 proton; the enhancement observed in this experiment was, in fact, 27%. Compound **18** is free to undergo ring inversion, but actually exists entirely in the diequatorial conformation (*24*). The NOE found when H–12 was saturated and H–7 observed was 19%. The fact that this enhancement is much less than the 27% observed for compound **17** might be due to one or more of the following: more efficient relaxation of H–7 by the methoxy protons in **18** than by the methylene proton in **17**; a different conformation for the two molecules so that the distance between H–7 and H–12 is greater in **18** than in **17**; or a difference in the preparation of the samples so that the result on **18** included a larger contribution from intermolecular relaxation than did the result on **17**. Colson *et al.* did not pursue this point.

The third compound (**19**) studied by Colson *et al.* undergoes ring inversion and so exists as a mixture of the axial and equatorial conformers. At $-45°C$ they found that saturation of the axial H–12 proton resonance resulted in a 27% enhancement of the H–7 resonance when the molecule was in the equatorial conformation and no enhancement of H–7 when the molecule was in the axial conformation, as would be expected. It might be noted that the kinetics of the ring inversion of this molecule could be studied very easily by the double-resonance techniques discussed in Chapter 7, although this has not yet been done. The problem would differ from those discussed in Chapter 7 in that the populations of the sites would be unequal and temperature dependent.

After examining these model systems, Colson *et al.* proceeded to apply the NOE to several problems in 7,12-dihydropleiadene stereochemistry. The first of these was the verification of the structures of the reactants and products in the base induced rearrangement shown in Fig. 8-6. The starting material for the rearrangement reaction was obtained by selective borohydride reduction of 8-methyl-7,12-pleiadione (**22**). Rearrangement of the starting material was accomplished by treating it with potassium *t*-butoxide in dimethylsulfoxide. The question to be resolved was whether the borohydride reduction produced **20** which then was rearranged to **21**, as drawn in Fig. 8-6, or whether the reduction really produced **21** which then rearranged to **20**. The identification was complicated by the fact that **20** and **21** were available only in small amounts and both exist as ketol-hemiketal mixtures in solution (*25*). However, in **21** H–12 is sufficiently close to the 11-methyl group so that an Overhauser effect should be observable between them, whereas in **20**, no NOE would be expected. When the 11-methyl group resonances were saturated in a solution of these

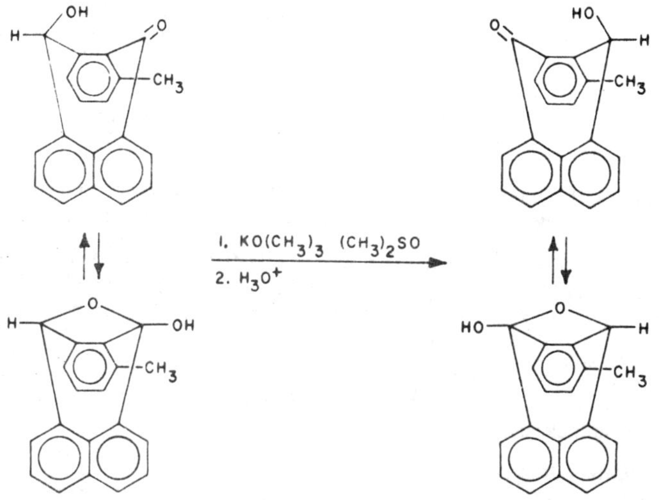

Fig. 8-6 Base induced rearrangement of the product obtained by borohydride reduction of 8-methyl-7, 12-pleiadione (**22**). [Reproduced by permission from J. G. Colson, P. T. Lansbury, and F. D. Saeva, *J. Amer. Chem. Soc.* **89**, 4987 (1967).]

compounds in dimethylsulfoxide-d_6, a 17% enhancement of the methine proton was observed in **21** and no enhancement was observed in **20**, thus confirming the structural assignments of Fig. 8-6.

Finally, Colson *et al.* applied the NOE to confirm their previous assignment of the NMR spectra of compounds **23** and **26** in which the low field methyl and methine signals had been arbitrarily assigned to the protons *syn* to the napthalene ring. This was done by performing Overhauser experiments on 7-methylene-8,11-dimethyl-7,12-dihydro-pleiadene (**27**). Saturation of the low-field methyl signal produced an enhancement of 20% in the upfield vinyl proton signal, confirming that these signals were due to the 8-methyl group and H–A, respectively. Saturation of the high field methyl produced a 16% increase in the intensity of the line due to the equatorial H–12 proton, H–B. The uncertainties in the assignments of the chemical shifts of the various protons were thus completely resolved.

These few examples should be sufficient to give the reader a feel for the way in which NOE experiments are interpreted in terms of molecular configuration. Therefore, the other examples of configuration analysis by NOE that have appeared in the literature will not be discussed in detail, but will only be summarized in the following pages.

Compound	Irradiate	Observe	Percentage of enhancement	Reference
28	H–6	H–9	25 ± 3	26
	CH$_3$–18	H–15	26 ± 3	

Dehydrovoachalotine

Experimental: A 10% solution of **28** in CDCl$_3$ with TMS added was studied at 100 MHz. The sample was degassed under vacuum and sealed under helium.

Conclusions: The NOE experiments confirmed that H–9 is the deshielded aromatic proton, and that the conformation of the ethylidene sidechain is CH$_3$–18, H–15 cis, as drawn.

Compound	Irradiate	Observe	Percentage of enhancement	Reference
29	*t*-butyl protons at 1 and 4	H–8 and H–5	15	27

1,4-di-*t*-butylnapthalene

Experimental: Not discussed

Conclusions: None

Compound	Irradiate	Observe	Percentage of enhancement	Reference
30	5-methyl	H–10	10	28

Experimental: Studies were conducted at 100 MHz on solutions of the compound in C$_6$D$_6$.

Conclusions: The 5-methyl proton group and H–10 proton must be on the same side of the ring. This was part of a study in which the structures of the sesquiterpenoid Bakkenolides-B, C, and D were determined.

Compound	Irradiate	Observe	Percentage of enhancement	Reference
31	H–7α	H–2	0	29
	H–9α	H–14	5*	
	H–9α	H–17	0	
	H–9α	H–18	5*	
	H–9β	H–14	5*	
	H–9β	H–17	0	
	H–9β	H–18	5*	
	H–10β	H–14	7	
Futoenone	H–10β	H–17	0	
	H–10β	H–18	7	
32	H–11α	H–14	17*	
	H–11α	H–17	0	
	H–11α	H–18	9*	
	H–14	H–9α	10*	
	H–14	H–9β	0*	
	H–14	H–10β	10*	
	H–14	H–11α	10*	
	H–18	H–9α	10*	
	H–18	H–9β	0*	
	H–18	H–10β	10*	
	H–18	H–11α	10*	

Compound	Irradiate	Observe	Percentage of enhancement	Reference

Experimental: The experiments on compounds **31** and **32** were performed at 100 MHz in mixtures of C_6D_6 and $CDCl_3$, with C_6D_6 being the predominant component. Exact concentrations were not specified in the report.

Conclusions: None. Since much of the spectrum is tightly coupled, the interpretation of the NOE results would be uncertain. In addition, the results marked by an asterisk are the percent increase in peak height rather than the percent increase in area and thus may not represent an Overhauser effect at all. The results on compounds **31** and **32** were essentially identical, and the discussion applies equally to both sets of data.

Compound	Irradiate	Observe	Percentage of enhancement	Reference
33	1-methyl	H–10	20	*30*

Zeylanane

Experimental: $CDCl_3$ was used as the solvent. No other details were given.

Conclusion: The NOE observed was taken as confirmation of the 1-methyl, H–10 cis configuration for this molecule.

Compound	Irradiate	Observe	Percentage of enhancement	Reference
34	H–9	H–1	25	*31, 32*
	2-methoxy	H–1	28	
	H–5	H–4	17	
	H–5′	H–4	17	
	3-methoxy	H–4	21	
	N-methyl	H–9	15	
	H–12	H–13	16	
35	H–9	H–1	22	
	N–CH$_3$	H–9	7	
	H–5	H–4	25	
	H–10 and N–CH$_3$	H–9	16	
Fumariline	H–9	H–10	19	
36	H–9	H–1	19	
	2-OCH$_3$	H–1	23	
	H–5	H–4	12	
	3-OCH$_3$	H–4	21	
	N–CH$_3$	H–9	7	
Fumaricine	H–10 and N–CH$_3$	H–10	20	

Experimental: The NOE's on compounds **34–36** were measured in $CDCl_3$ with TMS added for field-frequency locking.

Conclusions: The methoxy resonances in **34** were assigned on the basis of the NOE results. The NOE's also show that H–1 and H–9A are close together in **34** so that its conformation is as drawn. The NOE's in **35** and **36** are very similar to those in **34**, with the exception of an H–9, H–10 interaction which placed the methylenedioxy function unequivocally at the 12, 13 position in **35** and **36**. The methoxyl groups in **36** are shown to be at positions 2 and 3, and the conformational similarity of **34–36** is also indicated.

Compound	Irradiate	Observe	Percentage of enhancement	Reference
37	H–9A	H–1	14	33
	2–OCH$_3$	H–1	25	
	H–5	H–4	9	
	H–5	H–4	16	
	3–OCH$_3$	H–4	25	
	N–CH$_3$	H–9B	8	
	H–15A	H–12	3	
	H–15B	H–12	−7	
	H–15A	H–13	−8	
	H–15B	H–13	24	
	H–15B	H–15A	40	
	N–CH$_3$	H–15A	2	
	N–CH$_3$ and H–15B	H–15A	42	
	H–15A	H–15B	40	
	H–13	H–15B	8	
	H–15A and H–13	H–15B	47	

Ochotensimine

Experimental: The experiments were performed at 100 MHz on a carefully degassed 0.17 *M* solution of **37** in CDCl$_3$.

Conclusions: The observed enhancements may be satisfactorily interpreted in terms of the conformation drawn. The main point of this paper, the existence of positive and negative enhancements in all-proton systems, has already been discussed in detail in Section 3,G.

38	N–CH$_3$	H–14	24	34
	N–CH$_3$	H–9	0	
	H–14	H–12 and H–13	4	
	H–14	H–1	0	
	H–14	H–4	0	
	H–9	H–1	13	
	H–9	H–4	0	
	H–9	H–12 and H–13	0	

Ochrobirine

Experimental: Compound **38** was studied at 100 MHz in a degassed solution in CDCl$_3$ with TMS added for field-frequency locking.

Conclusions: The spectral assignments made using chemical shift and decoupling data were confirmed. The configuration of the molecule was determined to be as shown in the figure. Note that details such as the −OH groups being trans with respect to one another would have been difficult to determine by other methods.

39	N–CH$_3$	C–14	25	35

Sibiricine

Compound	Irradiate	Observe	Percentage of enhancement	Reference

Experimental: NOE studies on **39** were done at 100 MHz on a degassed solution in $CDCl_3$ with TMS added for field-frequency locking.

Conclusions: The configuration at C–14 is as drawn.

Compound	Irradiate	Observe	Percentage of enhancement	Reference
40 Fumaritine	3–OCH_3	H–4	24	*32*

Experimental: The NOE studies of **40** were done at 100 MHz on a degassed solution in $CDCl_3$ with TMS added for field-frequency locking.

Conclusions: The NOE data determined that the methoxy group was on C–3 rather than C–2, and thus showed that the –OH group was at C–2.

Compound	Irradiate	Observe	Percentage of enhancement	Reference
41 Polyoximic Acid	6–CH_3	H–4 and H–4′	22	*36*
	6–CH_3	H–5 and H–2	0	

Experimental: NOE experiments were done at 60 MHz on a 10% sample of **41** in D_2O with TMS added for field-frequency locking.

Conclusions: The 6-methyl group is trans with respect to the carboxyl group.

Compound	Irradiate	Observe	Percentage of enhancement	Reference
42	H–5	H–3	10	*37*
	H–1	H–7	5	
Pentalenolactone	H–14	H–7	15	
43				

Experimental: NOE experiments were done at 100 MHz in solution in $CDCl_3$.

Conclusions: H–3 and H–5 are close together, as are H–7 and H–14. The structure is thus **42** rather than **43**.

Compound	Irradiate	Observe	Percentage of enhancement	Reference
44	**44a**　$X = \beta$–OH, α–H; R = –CO–CH_3			*38*
	H–1	H–11	25	
	44b　$X = \beta$–CO–CH_3, α–H; R = –CO–CH_3			
Nagilactone C derivatives	18–CH_3	H–3α	17	
45	18–CH_3	H–3α	11	
	H–1	H–11	13	
3-Acetonagilactone D				

Experimental: Not reported

Compound	Irradiate	Observe	Percentage of enhancement	Reference

Conclusions: The NOE results were used to establish the configurations of Nagilactone C and Nagilactone D. In particular, the 18–CH$_3$, H–3 NOE required that H–3 be α, and H–1 and H–11 do not approach one another in any configuration except the one drawn.

Compound	Irradiate	Observe	Percentage of enhancement	Reference
46	9–H	H–2α	10	*39, 40*
	9–H	H–3α	0	
20,22-acetonide-2,3-diacetate-ponasterone A				

Experimental: Not reported

Conclusions: None

Compound	Irradiate	Observe	Percentage of enhancement	Reference
47	OCH$_3$	H–4	21	*41*
	OCH$_3$	H–1,6,7, and 8	0	
	H–4	OCH$_3$	7	
2-Hydroxy-3-methoxy-5-nitronapthalene	H–4	H–1,6,7, and 8	0	

Experimental: The experiments were done at 100 MHz on a 10% solution of **47** in acetonitrile with TMS added for field-frequency locking.

Conclusions: The structure was determined to be 2-hydroxy-3-methoxy rather than 3-methoxy-2-hydroxy or some other related compound.

Compound	Irradiate	Observe	Percentage of enhancement	Reference
48	–OCH$_3$	H–1	27	*41*
	–OCH$_3$	H–4	36	
	H–1	–OCH$_3$	4	
	H–4	–OCH$_3$	6	
2,3-dimethoxy-5-nitronapthalene				

Experimental: The experiments were done at 100 MHz on a degassed solution in acetonitrile, presumably with TMS added for field-frequency locking.

Conclusions: None

Compound	Irradiate	Observe	Percentage of enhancement	Reference
49	20–CH$_3$	H–11	7	*42*
	H–1 axial	H–11	7	
	H–1 equat.	H–11	17	
	H–9	H–11	18	
	H–9 and H–1 equat.	H–11	37	
Methyl 11α-bromo-12-oxopodocarpan-19-oate				

Experimental: NOE experiments were done at 100 MHz on a degassed 0.16 *M* solution of **49** in CD$_3$–CO–CD$_3$. The saturating rf field was on the order of 0.75–1.0 mG. Area integrations were performed at least 25 times for each signal. The mean areas observed were reproducible to within 1%.

Compound	Irradiate	Observe	Percentage of enhancement	Reference

Conclusions: **49** is in the 11 α-configuration, confirming the original assignment of Bible and Burtner (*43*). Note that the values for the NOE resulting from saturation of H–1 axial and H–1 equatorial are probably incorrect as these spins are tightly coupled, making it impossible to saturate the resonance of one without perturbing the populations which give rise to the other resonance.

Compound	Irradiate	Observe	Percentage of enhancement	Reference
50	5–OCH$_3$	H–3	0	*44*
	5–OCH$_3$	H–4	11	
	5–OCH$_3$	H–6	16	
	5–OCH$_3$	H–7	0	
	5–OCH$_3$	H–12	0	
	5–OCH$_3$	H–13 *cis*	0	
	5–OCH$_3$	H–13 *trans*	0	
	16 and 17–CH$_3$	H–3	0	
	16 and 17–CH$_3$	H–4	0	
	16 and 17–CH$_3$	H–6	0	
	16 and 17–CH$_3$	H–7	22	
	16 and 17–CH$_3$	H–12	0	
	16 and 17–CH$_3$	H–13 *cis*	0	
	16 and 17–CH$_3$	H–13 *trans*	0	
	14 and 15–CH$_3$	H–3	0	
	14 and 15–CH$_3$	H–4	0	
	14 and 15–CH$_3$	H–6	0	
	14 and 15–CH$_3$	H–7	0	
	14 and 15–CH$_3$	H–12	12	
	14 and 15–CH$_3$	H–13 *cis*	25	
	14 and 15–CH$_3$	H–13 *trans*	0	

Poncitrin

Experimental: The NOE experiments were done at 100 MHz on 10% (w/v) degassed solutions of **50** in CDCl$_3$ with TMS added for field-frequency locking.

Conclusions: The structure drawn is the only one consistent with the NOE results. In particular, the enhancement of H–4 and H–6 when the methoxy group was saturated fixed the methoxy at position 5. The NOE results also confirmed the presence of the 2,2-dimethylchromene and 1,1-dimethylallyl groups, and the assignment of their *gem*-dimethyl groups.

Compound	Irradiate	Observe	Percentage of enhancement	Reference
51	5–CH$_3$	H–10	30	*45*

Portentol

Experimental: The NOE was observed at 100 MHz on a solution in CDCl$_3$.

Conclusions: The NOE results plus other data established that the methyl groups on the tetrahydropyran ring are all equatorial.

Compound	Irradiate	Observe	Percentage of enhancement	Reference
52	3–CH$_3$ axial	H–8 axial	15	*46*
	8–N–CH$_3$	H–8 axial	15	
	1–N–CH$_3$	H–8 axial	0	

Physostigmine

Experimental: The NOE studies were made at 100 MHz on a 10% solution of **52** in deuterochloroform with TMS added for field-frequency lock.

Conclusions: The fused, five-member rings are cis. Further, the 8–N–methyl is trans, although rapid inversion at the 1–nitrogen could result in a large enough mean distance between the 1–N–methyl and H–8 to give the zero enhancement observed.

Compound	Irradiate	Observe	Percentage of enhancement	Reference
53	H–3′	H–1′	5 ± 4	*47*
	H–3′	H–8	0	
	H–2′	H–1′	10 ± 5	
	H–2′	H–8	0	
	H–1′	H–2′	10	
	H–1′	H–8	36 ± 2	
	H–8	H–2′	0	
	H–8	H–1′	32	

2′, 3′-isopropylidene-
  3,5′-cycloguanosine

Experimental: The NOE studies were made at 100 MHz on a degassed 0.25 *M* sample in dimethylsulfoxide-d_6. The probe temperature was maintained at 31 ± 1°C. Field-frequency locking was accomplished by using a coaxial tube assembly with hexamethyldisilizane in the annular space, thus eliminating intermolecular relaxation due to the lock material. Spectra were recorded using a 50 Hz sweep width and 1.0 Hz/sec sweep rate, and integrated by planimeter.

Conclusions: See Section 3,H for a thorough discussion of these experiments.

Compound	Irradiate	Observe	Percentage of enhancement	Reference
54	21–CH$_3$	H–22	13	*48*

Experimental: Not reported.

Conclusions: It was concluded that the 21-methyl group was cis to H–22. While probably correct, comparison of the reported result with the result of the experiment in which H–20 is saturated while H–22 is observed would be required in order to make it certain that the 21-methyl is closer to H–22 than is H–20.

Compound	Irradiate	Observe	Percentage of enhancement	Reference
55	H–9α	H–2	7	*49*

2,3,22,24-Ponasterone C
  Tetraacetate

Experimental: Not reported.

Conclusion: The NOE was used along with other data to demonstrate that there is a hydroxy group at the 5-position and that the hydroxy is in the β configuration.

Compound	Irradiate	Observe	Percentage of enhancement	Reference
56 *cis*-Theaspirone	$5-CH_3$ $5-CH_3$	H–1 H–6	10 20	*50*
57 *trans*-Theaspirone	$5-CH_3$ $5-CH_3$	H–1 H–6	−1 10	

Experimental: The experiments were done at 100 MHz on degassed samples in $CDCl_3$ with TMS added for field-frequency locking. The experiments were done at about 24°C.

Conclusions: The NOE results confirmed the previous assignment (*51, 52*) of the cis and trans geometries to these isomers. The different value obtained for the $5-CH_3$, H–6 NOE in the two isomers was not discussed, but may have been due to differences in sample preparation resulting in different intermolecular contributions to the relaxation of H–6.

58	H_0	H–1	8	*53a*
	H_0	H–2	−5	
	H–1	H–2	60 (47)*	
	H–1	H_0	23	
	H–2	H–1	40	
	H–2	H_0	−5	

1,11-Dimethyl-
  5,7-dihydrodibenz
  (c,e) thiepin

Experimental: A sample was prepared which was deuterated in the *meta* position of both rings. The NOE experiments were then done at 100 MHz on a degassed sample of less than 5% concentration (w/v). The solvent was probably dimethylsulfoxide-d_6. The correctness of the 60% enhancement found in the experiment marked by an asterisk was questionable, so the sample was submitted to Bell and Saunders who repeated the experiments. The latter workers found only a 47% enhancement for the experiment in question, but both laboratories were in agreement on the other results. The reason for the discrepancy is not known.

Conclusions: H–1 is closer to H_0 than are the other protons, H–2—H–4, and thus H–1 has the *pro*-S configuration in the R-biphenyl. H–1 gives rise to the lowest field line of the protons 1–4.

59	deuterated at the *meta* position			*53a*
	H_0	H–1	10	
	H_0	H–2	−1	
	H–1	H–2	58 (48)*	
	H–1	H_0	21	
	H–2	H–1	30	
	H–2	H_0	0	

Compound	Irradiate	Observe	Percentage of enhancement	Reference
	deuterated at the *meta* and 1 positions			
	H_0	H–2	6	
	H–2	H_0	0	
	deuterated at the 2 position			
	H_0	H–1	0	
	H–1	H_0	23	

1,11-dimethyl-
  5,7-dihydrodibenz
  (c,e) thiepin-S-dioxide

Experimental: The experiments were done at 100 MHz on degassed samples containing less than 5% (w/v) of **59** in dimethylsulfoxide-d_6. As with the experiments on **58**, the value marked with an asterisk was considered unreasonable and was checked by Bell and Saunders on the same sample. Bell and Saunders obtained the value in parentheses for the NOE in question, but agreed with the other results.

Conclusions: The lowest field line of the benzylic protons (H–1) has the "*pro*-S in R" configuration.

Compound	Irradiate	Observe	Percentage of enhancement	Reference
60	H–1 and H_m deuterated—$(CD_3)_2SO$			*53a*
	H_0	H–4	5	
	H–4	H_0	3	
	1–CH_3	H_p	28	
	1–CH_3	$H_p{}'$	14	
	2–CH_3	H_p	9	
	2–CH_3	$H_p{}'$	24	
	H–2, H–3, H–4, and H_m deuterated—$(CD_3)_2SO$			
	H_0	H–1	31	
	H–1	H_0	33	
	H–1 and H_m deuterated—$CDCl_3$			
	H_0	H–4	0	
	H–2, H–3, H–4, and H_m deuterated—$CDCl_3$			
	H_0	H–1	0	
	H–1	H_0	0	

Compound	Irradiate	Observe	Percentage of enhancement	Reference
		H_m deuterated—$CDCl_3$		
	H–1	H–4	28	
	H–4	H–1	24	
	1–CH_3	H_p	10	
	1–CH_3	$H_p{}'$	16	
	2–CH_3	H_p	11	
	2–CH_3	$H_p{}'$	10	

Experimental: The NOE experiments were performed at 100 MHz on degassed samples having a concentration of less than 5% (w/v). The solvent was either $CDCl_3$ or $(CD_3)_2SO$, as indicated.

Conclusions: The results establish a *pro*-S configuration for H–1 and a *pro*-R configuration for H–4 in the R-biphenyl. The low results found in $CDCl_3$ may be due in part to dimerization (*53b,c*). The nonzero values for the enhancements of H_p and $H_p{}'$ upon saturating either methyl resonance probably indicates that both methyls were at least partially saturated during these experiments. Table 1 of Fraser and Schuber (*53a*) gives the chemical shift between the methyls as 0.026 ppm in $CDCl_3$ (0.020 for the *m*-deutero derivative in $CDCl_3$), and 0.05 ppm for the *m*-deutero compound in $(CD_3)_2SO$. It is very unlikely that resonances this close together could be selectively saturated.

Compound	Irradiate	Observe	Percentage of enhancement	Reference
61	CH_3–A	cyclobutyl	24	*54*
	CH_3–B	cyclobutyl	0	
	CH_3–C	cyclobutyl	0	
	CH_3–A	cyclopropyl	0	
	CH_3–B	cyclopropyl	18	
	CH_3–C	cyclopropyl	0	
62	CH_3–A	cyclobutyl	28	
	CH_3–B	cyclobutyl	32	
	CH_3–C	cyclobutyl	11	
	CH_3–A	cyclopropyl	8	
	CH_3–B	cyclopropyl	14	
	CH_3–C	cyclopropyl	20	

Experimental: Compound **61** was studied at 100 MHz in an approximately 0.2 M solution in $CDCl_3$ with TMS added for field-frequency locking. Compound **62** was presumably studied under similar conditions, but this was not made explicit in the report. The label "cyclobutyl" in the table refers to the two methine protons in the cyclobutyl ring. "Cyclopropyl" refers to the four bridgehead protons in the cyclopropyl groups. The cyclobutyl protons give rise to a singlet in the spectrum, while the cyclopropyl protons give rise to an AB quartet.

Conclusions: The NOE studies allowed assignment of the structures **61** and **62** to the isomers produced by photodimerization of 3-carene-2,5-dione. An alternate configuration for **62** was also ruled out on the basis of the NOE experiments.

C. Determination of Molecular Conformation

The range of conformations which a molecule may assume is limited by bond lengths, the order in which the atoms are connected together, the configuration at each atom, and the size and shape of the groups involved. A study of molecular conformation is an attempt to determine which of the conformations allowed by these constraints are actually assumed by the molecule. As far as the NOE experiment is concerned, the difference between a study of configuration and of conformation vanishes when only a single conformation actually exists; thus the separation of the material in this chapter into applications of the NOE in configuration and in conformation studies is somewhat academic because in all cases reported in the literature it has been assumed that only a single conformation does in fact occur. It is nevertheless necessary to keep this distinction in mind because the existence of several conformations, or a range of conformations, will result in the experimental enhancements being averages over the geometries present, as was discussed in Chapter 4. Failure to take this averaging into account can lead to erroneous conclusions. There is, in fact, a possibility that some of the studies reviewed in this chapter are in error for this very reason. In the absence of additional evidence, we will not comment further on this possibility.

The first application of the Overhauser effect to conformation analysis was made by Woods *et al.* (55) as part of their work on the structures of the ginkgolides. The enhancements observed for the ginkgolides GA (**63**), GB (**64**), GC (**65**), and several derivatives of GC (**66, 67**) in trifluoroacetic acid are presented in Table 8-2.

TABLE 8-2

NUCLEAR OVERHAUSER EFFECTS IN SEVERAL GINKGOLIDES
AND GINKGOLIDE DERIVATIVES[a]

		Compound					
Irradiate	Observe	63	64	65	66	67	68
t-butyl	*I*	30	25	20	33	27	30
t-butyl	*J*	22	21	20	22	0	0
t-butyl	*F*	—	—	4	16	19	24
t-butyl	*E*	—	—	—	6	13	14
t-butyl	*H*	0	0	0	0	6	12
E	*J*	—	—	0	0	10	7
J	*E*	—	—	0	0	23	20

[a] Data from Woods *et al.* (55).

In **63–66** significant enhancements of the signals from protons I, J, E, and F were observed when the *t*-butyl resonance was saturated. The intensities of other lines in the spectrum were unaffected. In several of the cases where the values for f_E(*t*-butyl) and f_F(*t*-butyl) are missing in the table, increases in the heights of the E and F resonances were observed but the enhancement could not be measured accurately due to overlap of other resonances or the proximity of the observed to the saturated line. As these experiments were done in trifluoroacetic acid and the fluorine nucleus has a large magnetic moment, it would be expected that the intermolecular dipole–dipole relaxation was large. Thus the large enhancements observed indicated that the *t*-butyl protons were very close to I, J, E, and F, and that the *t*-butyl group blocked the latter protons from close approach by solvent molecules. When the B ring is in the conformation drawn with the *t*-butyl in a quasi-equatorial position, the *t*-butyl is in fact very close to all four of the protons, I, J, E, and F; this conformation was therefore assigned to **63–66**.

The situation was found to be quite different in the two *iso*-GC derivatives, **67** and **68**. In these compounds the resonance of proton J was not enhanced upon saturating the *t*-butyl resonance, but an enhancement was observed for proton H. These changes in the NOE's correlate well with a shift of the lactone ring from C–6 to C–7 as had been proposed on the basis of other evidence (*56a, b, c*). The shift of the position of the lactone results in a change in the conformation of the B ring so that the *t*-butyl occupies a *quasi*-axial position as is reflected in the NOE data.

TABLE 8-3

THE NUCLEAR OVERHAUSER EFFECT IN ZEYLANINE (69)[a]

		Irradiate					
Solvent	Observe	H–6	H–7β	H–10	H–13	H–14	–OCOCH$_3$
CDCl$_3$	H–2α	2	0	0	—	0	—
	H–6	—	11	8	0	2	—
	H–7β	18	—	–4	13	0	—
	H–10	6	—	—	—	20	—
	H–12	—	0	0	19	0	—
C$_6$D$_6$	H–2α	0	—	0	—	0	0
	H–6	—	14	—	0	6	0
	H–7β	14	—	3	8	0	0
	H–10	—	–2	—	—	17	0
	H–12	0	—	0	11	—	0

[a] Data from Tori *et al.* (*57*).

TABLE 8-4

THE NUCLEAR OVERHAUSER EFFECT IN ZEYLANANE (70)[a]

Solvent	Observe	Irradiate					
		H–6	H–7β	H–10	H–13	H–14	–OCOCH$_3$
CDCl$_3$	H–2α	0	—	0	—	0	—
	H–6	—	15	10	—	0	—
	H–7β	14	—	–4	17	0	—
	H–10	8	—	—	—	20	—
	H–12	—	—	—	16	—	—
C$_6$D$_6$	H–2α	0	0	–3	—	0	–3
	H–6	—	14	12	–3	0	–4
	H–7β	12	—	–2	14	—	0
	H–10	10	–4	—	—	19	0
	H–12	–3	–4	–2	15	—	0

[a] Data from Tori *et al.* (*57*).

The conformations of zeylanine (**69**) and zeylanane (**70**) have also been established by use of the NOE (*57*). The experiments were performed at 100 MHz on degassed solutions containing about 5% (w/v) of the compound in CDCl$_3$ and C$_6$D$_6$ with TMS added for field locking. The enhancements are recorded in Tables 8-3 and 8-4. In **69** the enhancement of H–6 when either H–14 or H–10 was saturated was taken to mean that these spins are relatively close to one another so that the ring must be in the conformation with H–6 and the 1-methyl *syn*. The enhancement of H–6 when H–7β was saturated indicated that these protons were cis. The negative enhancement of H–10 upon saturating H–7β would be expected because the distances between H–7β and H–6 and between H–10 and H–6 are both less than the distance between H–7β and H–10.

The NOE's observed on **70** are very similar to those on **69**, and the conclusions are essentially the same. The main differences are in the magnitudes of the enhancements: $f_6(14)$ is now zero while $f_{7\beta}(6)$ and $f_6(10)$ are larger and $f_{10}(7\beta)$ is more negative. These changes would indicate that forming the 5,6 epoxide shifts H–6 a little closer to H–7β and H–10.

In both **69** and **70** $f_{10}(14–CH_3)$ are quite large, as would be expected from the proximity of the spins and the fact that the separations between them are independent of molecular conformation. The other enhancements discussed are all quite small. Whether this was due to a large intermolecular contribution to spin relaxation, the geometry of the molecule, or averaging over several conformations cannot be decided on the basis of the data presented.

TABLE 8-5

NUCLEAR OVERHAUSER EFFECTS IN ISOLINDERALACTONE (71) IN $CDCl_3$[a]

| | Irradiate | | | | | | | | | |
Observe[b]	H–2c	H–3t	H–3c	H–5	H–6	H–9α	H–9β	H–12	H–13	H–14
H–1	—	—	0	6	—	12	12	—	—	2
H–2t	—	—	0	0	—	0	12	—	—	−4
H–2c	—	—	0	0	—	0	0	—	—	20
H–3t	0	—	37	0	—	—	0	—	0	0
H–3c	0	46	—	0	0	—	—	—	0	0
H–5	—	8	−5	—	12	—	—	—	−5	0
H–6	—	—	0	9	—	—	—	0	7	0
H–9α	—	—	—	—	—	—	—	—	—	2
H–9β	—	—	—	—	—	—	—	—	—	0
H–12	—	0	0	0	−5	—	—	—	23	0
10–CH$_3$	—	0	—	0	—	0	0	—	—	—

[a] Data from Tori and Horibe (58).
[b] c and t label cis and trans, respectively.

TABLE 8-6

NUCLEAR OVERHAUSER EFFECTS IN ISOLINDERALACTONE (71) IN C_6D_6[a,b]

| | Irradiate | | | | | | | | | |
Observe[c]	H–2c	H–3t	H–3c	H–5	H–6	H–9[d]	H–12	H–13	H–14
H–1	—	—	0	—	—	—	—	—	0
H–2t	—	—	0	0	—	0	—	0	−4
H–2c	—	—	0	1	—	1	0	0	18
H–3t	—	—	36	0	—	0	—	—	0
H–3c	0	45	—	0	0	0	—	—	0
H–5	—	—	—	—	—	—	—	—	—
H–6	—	—	0	—	—	—	−1	7	0
H–9	—	—	—	—	—	—	—	—	—
H–12	—	0	0	0	−4	—	—	26	0
10–CH$_3$	—	0	—	0	—	0	—	—	—

[a] Data from Tori and Horibe (58).
[b] Several triple resonance experiments were also reported. These were: H–1 {H–5, H–9}, 30%; H–5 {H–3t, H–6}, 21%; and H–6 {H–5, H–13}, 16%.
[c] c and t label cis and trans, respectively.
[d] 9α and 9β are not resolved in this solvent.

Another study which made extensive use of the NOE was the determination of the conformation of isolinderalactone (**71**) by Tori and Horibe (*58*). An examination of models showed that there were four possible conformations for the cyclohexene nucleus: two with both H–9β and the 10–CH$_3$ axial, and the other two with both H–9β and H–10 axial. To determine which of these conformations actually occurred, Tori and Horibe performed NOE experiments at 100 MHz on samples in both CDCl$_3$ and C$_6$D$_6$. The results are summarized in Tables 8-5 and 8-6. The fact that the H–9, H–5, and H–3t resonances were not enhanced upon saturation of the 10-methyl resonance in addition to (1) long-range couplings of H–9 with H–6β, H–12, and H–14; and (2) a 0.3-Hz coupling between H–5β and H–14, demonstrated that the cyclohexene nucleus was in the half-chair conformation with 10–CH$_3$, H–9β, and H–5 oriented axially. Examination of the values for $f_1(5)$, $f_1(9\alpha)$, $f_1(9\beta)$, and $f_{2C}(14)$ also showed that the 10-vinyl group existed predominantly in the conformation with H–2C near the 10-methyl and H–1 between H–5 and the two H–9 protons.

Compound	Irradiate	Observe	Percentage of enhancement	Reference
72	15α–CH$_3$	H–2	10–20	59
	15α–CH$_3$	H–9		
	12–CH$_3$	H–21		
	12–CH$_3$	H–10	>0	
	H–7	H–10		

Taxinine

73   TB: R = O; R$_7$ = H

74   TE: R = CHOAc; R$_7$ = H

75   TJ: R = CHOAc; R$_7$ = OAc

Experimental: Not discussed.

Conclusions: The only structure for **72** which can account for all the observed enhancements is the structure drawn. The stereochemistry at C–5 remains uncertain, however. Compounds **73–75** showed enhancements similar to those in **72** and are thus in essentially the same conformation as **72**. A 10–15% enhancement of H–13 upon saturating the 15β–CH$_3$ was also observed in TE and TJ, so the 13-proton must be in the β-configuration in these molecules.

76	H–5	H–6	4–5	60
	15–CH$_3$	H–6	21	
	17–CH$_3$	H–6	10	
	18–CH$_3$	H–6	2–3	

Methyl 12-Acetoxy-
6-bromo-
7-oxopodacarpa-
8,11,13-trien-16-oate

Experimental: NOE data was obtained on a 0.21 *M* solution of **76** in CCl$_4$ and also on a 0.20 *M* in (CD$_3$)$_2$CO. In both cases approximately 5% C$_6$H$_6$ was added for field-frequency locking. The results were essentially identical in both solvents.

Conclusions: The enhancement of H–6 upon saturating the 17-methyl shows that the bromine is α. The fact that f_6(15–CH$_3$) > f_6(17–CH$_3$) implies that the 15–CH$_3$ is closer to H–6 than the 17–CH$_3$ is, so that ring B must be in a boat or twist-boat conformation. The remaining enhancements are consistent with this interpretation.

77	low field CH$_3$	H–3	20 ± 3	61
	high field CH$_3$	H–3	13 ± 3	

R = CH$_3$ or C$_6$H$_5$OCH$_2$–

Experimental: NOE experiments were performed at 100 MHz on samples that were less than ~5% (w/v) concentration in CDCl$_3$. TMS was added as internal lock material.

Compound	Irradiate	Observe	Percentage of enhancement	Reference

Conclusions: H–5 and H–6 were determined to be cis because of their large coupling constant, $J = 4$ Hz. Only H–5 is enhanced when the high-field methyl is saturated so this methyl is α. Since only H–3 is enhanced when the low field methyl is saturated, this is the β-methyl. These enhancements also indicated that the thiazolidine ring conformation of the (R)-sulfoxides is approximately the same as in the (S) sulfoxides, but quite different than in the analogous sulfides. The conformation is as drawn.

Compound	Irradiate	Observe	Percentage of enhancement	Reference
78		in solution in CCl₄		62
	low field CH₃	H–3	21 ± 3	
	low field CH₃	H–5	0 ± 3	
	high field CH₃	H–3	7 ± 3	
	high field CH₃	H–5	0 ± 3	
		in solution in C₆D₆		
	low field CH₃	H–3	22 ± 3	
	low field CH₃	H–5	0 ± 3	
	high field CH₃	H–3	12 ± 3	
	high field CH₃	H–5	0 ± 3	
Phenoxymethylpenicillin				
79		in solution in CCl₄		
	low field CH₃	H–3	26 ± 3	
	low field CH₃	H–5	0 ± 3	
	high field CH₃	H–3	0 ± 3	
	high field CH₃	H–5	14 ± 3	
		in solution in C₆D₆		
	low field CH₃	H–3	27 ± 3	
	low field CH₃	H–5	0 ± 3	
	high field CH₃	H–3	0 ± 3	
	high field CH₃	H–5	14 ± 3	
Phenoxymethylpenicillin sulfoxide				
80		in solution in CCl₄		
	2β–CH₃	H–3	22 ± 3	
	2β–CH₃	H–5	0 ± 3	
	2α–CH₃	H–3	0 ± 3	
	2α–CH₃	H–5	11 ± 3	

Compound	Irradiate	Observe	Percentage of enhancement	Reference
		in solution in C_6D_6		
	2β–CH$_3$	H–3	18 ± 3	
	2β–CH$_3$	H–5	0 ± 3	
	2α–CH$_3$	H–3	0 ± 3	
	2α–CH$_3$	H–5	10 ± 3	
Phenoxymethylpenicillin sulfone				

Experimental: The NOE experiments were performed at 100 MHz on nitrogen sparged solutions containing less than 5% (w/v) of the compound in CCl_4 or C_6D_6, as indicated. TMS was used for field locking in CCl_4 solutions, and C_6H_6 was used in the solutions in C_6D_6.

Conclusions: The low-field methyl resonances for compounds **78–80** correspond to the 2β-methyl protons (in both solvents). The enhancement of H–5 upon saturating the 2α-methyl resonance in **79** and **80**, but not in **78**, indicated that a different conformation is preferred by the sulfide than by the oxides (compare the drawings). The sulfide and sulfoxide assume the same conformation in solution as they do in their respective crystals.

Compound	Irradiate	Observe	Percentage of enhancement	Reference
81	2α–CH$_3$	H–3	14	*63*
	2α–CH$_3$	H–3	26	
Methyl Phthalimidopenicillinate				
82	2α–CH$_3$	H–3	0	
	2β–CH$_3$	H–3	18	
	2α–CH$_3$	H–5	~9	

Experimental: Not reported.

Conclusions: The assignment of the spectrum was based on the $2\alpha - CH_3$, H–5 NOE in compound **82**. The enhancements were interpreted as meaning the sulfide and sulfoxide are in different conformations, just as was found with phenoxymethylpenicillin (**79**) and its sulfoxide (**80**).

Compound	Irradiate	Observe	Percentage of enhancement	Reference
83		in solution in $CDCl_3$ or C_6D_6		*64*
	3–CH$_3$	H–2α	0	
	3–CH$_3$	H–2β	0	
		in solution in $(CD_3)_2SO$		
	3–CH$_3$	H–2	~5	
Δ^3-Deacetoxycephalosporin Sulfoxide				

Compound	Irradiate	Observe	Percentage of enhancement	Reference

Experimental: The NOE experiments were performed at 100 MHz on solutions of less than 5% concentration (w/v) in the solvent indicated. TMS was added for field locking.

Conclusions: The two protons at C–2 give rise to an AB quartet in $CDCl_3$ and C_6D_6 and it was hoped that the NOE experiments would allow assignment of the lines of this quartet to transitions of the 2α or 2β proton. No decision was possible because of the unexpected absence of an NOE. The zero enhancement may have been due to the fact that the mutual relaxation of H–2α and H–2β is much more efficient than the relaxation of either spin by the 3-methyl group as suggested by Cooper *et al.* (*64*); or it may have been the result of the relative positions of the three groups of spins (see Fig. 4–1).

Compound	Irradiate	Observe	Percentage of enhancement	Reference
84a $R^3 = R^2 =$ lone pair	3–CH_3	H–2	17	*64*
	3–CH_3	H–4	12	
84b $R^3 =$ lone pair	3–CH_3	H–2	12	
$R^2 = O$	3–CH_3	H–4	10	
84c $R^3 = O$				
$R^2 =$ lone pair	3–CH_3	H–2	12	

Experimental: The NOE studies were performed at 100 MHz on solutions of less than 5% (w/v) concentration. Other conditions were not reported.

Conclusions: The similarity of the long-range coupling constants and of the Overhauser effects observed in these three compounds indicated a uniformity of conformation among them. The conformation drawn agreed best with all the data presented.

Compound	Irradiate	Observe	Percentage of enhancement	Reference
85	4–CH_3	H–6	15	*65*
	4–CH_3	H–2	10	
	4–CH_3	H–1	0	
	10–CH_3	H–2	15	
	10–CH_3	H–5	0	

Dihydrotamaulipin-A
   Acetate

Experimental: The NOE data was collected at 100 MHz on a solution of **85** in C_6D_6.

Conclusions: The relatively large enhancements between H–6, H–2, and the 4–CH_3 group imply that these spins are close together. The 4–CH_3 group has the same direction in space as H–6β and H–2. The NOE's between H–2 and both the 4- and 10-methyls also require that these methyls be *syn*. Therefore the conformation of the molecule is as drawn.

Compound	Irradiate	Observe	Percentage of enhancement	Reference
86	11–CH_3	H–12	22 ± 2	*66*
	11–CH_3	H–6β	6 ± 2	
	5–CH_3	H–6β	28 ± 2	

Furanoeremophilan-14β,
   6α-olide

Compound	Irradiate	Observe	Percentage of enhancement	Reference
87	11–CH$_3$	H–12	17$\pm$2	
	11–CH$_3$	H–6β	7$\pm$2	
	5–CH$_3$	H–6β	23$\pm$2	
	5–CH$_3$	H–14β	8$\pm$2	
	5–CH$_3$	H–14α	$-2\pm$2	

Experimental: The NOE studies of **86** and **87** were done at 100 MHz on degassed solutions in CDCl$_3$. About 5% (w/v) TMS was added for field-frequency locking.

Conclusions: H–6 and C–CH$_3$ are cis in **86** as is shown by the large NOE between these spins. The NOE between H–6 and the 11β–CH$_3$ on the furan ring indicates that C–6 is attached to the other β-position of the furan ring. The NOE studies plus other data led to the structure drawn for **86**. The similarity of the results obtained on **86** and **87** showed that there was no stereochemical change in converting **86** to **87**.

Compound	Irradiate	Observe	Percentage of enhancement	Reference
88	19–CH$_3$	H–6(A$_1$)	45	67
	19–CH$_3$	H–6(A$_2$)	35	
	19–CH$_3$	H–7(B$_1$)	18	
	19–CH$_3$	H–7(B$_2$)	15	
	19–CH$_3$	H–9	0	
	19–CH$_3$	H–3	35	
	9–H	H–6	10	
	9–H	H–7	0	
	9–H	H–3	0	

trans-tachysterol-3,5-
    dinitro-4-methylbenzoate

Experimental: 60 mg of **88** was dissolved in 0.5 ml of CDCl$_3$, and a few drops of hexamethyldisiloxane were added for field-frequency locking, and the sample was degassed. The NOE studies were then done at 100 MHz.

Conclusions: Comparing the enhancements of H–6 and H–7 upon saturating H–9 strongly supports the assignment of the S$_{5,6}$ trans, S$_{7,8}$ cis conformation for **88** as opposed to the S$_{5,6}$ trans, S$_{7,8}$ trans conformation. The enhancement of H–7 upon saturating the 19–CH$_3$ resonance is probably a result of the tight coupling of H–6 and H–7 (hence the AB notation for these experiments in the table) or partial saturation of another resonance. The large enhancement of H–3 upon saturating the 19–CH$_3$ resonances may have been due to partially saturating the H–2 resonance. The value of f_3(19–CH$_3$) is suspiciously large, however, even after allowing for the possible saturation of H–2.

Compound	Irradiate	Observe	Percentage of enhancement	Reference
89	H–4	H–11	16	68
	H–11	H–4	23	

cis-4,5,7,7-tetrabromo-
    2,3-benzocycloheptenone

Compound	Irradiate	Observe	Percentage of enhancement	Reference
90	H–4	H–11	17	
	H–11	H–4	20	

trans-4,5,7,7-tetrabromo-
    2,3-benzocycloheptenone

Compound	Irradiate	Observe	Percentage of enhancement	Reference

Experimental: The NOE experiments were done at 100 MHz on degassed solutions of **89** and **90** in $CDCl_3$ with TMS added for locking.

Conclusions: Coupling constants, chemical shifts, and Overhauser effects were used to show that **89** and **90** have the same planar structure. Chemical shifts and coupling constants were then used to assign the conformations drawn.

Compound	Irradiate	Observe	Percentage of enhancement	Reference
91	–OH	H–2	~26	68
	–CHO	H–2	~8	
	–CHO and –OH	H–2	~34	
	H–2	–CHO	~8	
	–OCH$_3$	–CHO	~24	
	–OH	H–8	0	
	–OH	H–5	0	

Experimental: The NOE experiments were done at 100 MHz on samples in $CDCl_3$ with TMS added for field-frequency locking.

Conclusions: The NOE's rigorously determined the structure of **91** to be as drawn.

Compound	Irradiate	Observe	Percentage of enhancement	Reference
92	high field CH$_3$	H–7, *endo*	45	69
	low field CH$_3$	H–6	35	

3-carbomethoxy-5,5-
    dimethyl-4-thia-1-aza-
    bicyclo [4.1.0]
    hept-2-ene

Experimental: Not reported.

Conclusions: The NOE results indicated that the axial 5-methyl group was in close proximity to the *endo* H–7 proton while the equatorial 5–CH$_3$ was close to H–6, and thus the conformation was deduced to be as drawn.

Compound	Irradiate	Observe	Percentage of enhancement	Reference
93	H–8	H–1′	26	70
	H–1′	H–8	39	
	H–2′ and H–3′	H–8	1	

2′,3′-isopropylidene-
    3,5′-cycloguanosine

Compound	Irradiate	Observe	Percentage of enhancement	Reference
94	H–8	H–1′	20	
	H–1′	H–8	12	
	H–2′ and H–3′	H–8	12	
	H–5′ and H–5″	H–8	10	
	H–4′	H–8	2	

2′,3′-isopropylideneguanosine

Compound	Irradiate	Observe	Percentage of enhancement	Reference
95	H–1′	H–8	20	
	H–2′	H–8	12	
	H–3′	H–8	12	
	H–5′ and H–5″	H–8	10	
	H–1′	H–2	2	

2′,3′-isopropylideneadenosine

Compound	Irradiate	Observe	Percentage of enhancement	Reference

Experimental: The experiments were performed at 100 MHz on degassed samples in $(CD_3)_2SO$. The samples were prepared in coaxial tubes with hexamethyldisilizane in the annular space for field locking. The concentrations of **93, 94,** and **95** were 0.27 M, 0.44 M, and 0.58 M, respectively.

Conclusions: The data obtained for **93** shows the pattern of enhancements that is expected for a purine nucleoside locked into the *syn* conformation. Comparison of the NOE's for the three nucleosides indicated that both *syn*-like and *anti*-like conformations are allowed for both **94** and **95**, but that the *anti* range is less favored in **95** than in **94**—a difference that had not been anticipated [cf also (*47*)].

Compound	Irradiate	Observe	Percentage of enhancement	Reference
96	~0.25 M in $(CD_3)_2SO$			*71*
	H–2', H–3', H–4'	H–6	18	
	~0.25 M in 1:1 $(CD_3)_2SO:C_6D_6$			
	H–1'	H–6	8	
	H–2'	H–6	13	
	H–3', H–4', H–5'	H–6	4	
	~0.25 M in D_2O			
	H–1'	H–6	13	
	H–2'	H–6	14	
	H–3'	H–6	10	
	H–4', H–5'	H–6	0	
Cytidine				
97	~0.25 M in $(CD_3)_2SO$			
	H–1'	H–6	10	
	H–2' and H–3'	H–6	13	
	H–5'	H–6	10	
	~0.25 M in D_2O			
	H–5'	H–6	5	
	~0.25 M in 3:1 $(CD_3)_2SO:C_5H_5N$			
	H–1'	H–6	6	
	H–2', H–3', H–4'	H–6	18	
	H–5'	H–6	12	
Uridine				

Compound	Irradiate	Observe	Percentage of enhancement	Reference
98		$\sim$0.25 M in $(CD_3)_2SO$		
	H–1′	H–6	19	
	H–2′	H–6	14	
	H–3′	H–6	4	
	H–5′	H–6	5	
		$\sim$0.25 M in 3:1 $(CD_3)_2SO:D_2O$		
	H–1′	H–6	19	
	H–2′	H–6	10	
	H–3′	H–6	4	
2′,3′-isopropylidene-uridine				
99		$\sim$0.25 M in $(CD_3)_2SO$		
	H–1′	H–6	7	
	H–2′	H–6	18	
	H–3′	H–6	8	
	H–5′	H–6	4	
		$\sim$0.25 M in 4:1 $D_2O:(CD_3)_2SO$		
	H–1′	H–6	5	
	H–2′	H–6	16	
	H–3′	H–6	8	
	H–5′	H–6	4	
Thymidine				

Experimental: The experiments were performed at 100 MHz on $\sim$0.25 M solutions prepared in the solvents indicated in the table. Several mixed solvent systems were used in order to obtain a suitable compromise between the solubility of the compound and the chemical shifts separating the lines in its spectrum. The solutions were degassed and sealed in a coaxial tube assembly containing hexamethyldisilizane in the annular space for field locking.

Conclusions: The significant enhancement of H–6 upon saturating H–1′ in all four of the nucleosides was interpreted as meaning that conformations in the *syn* as well in the *anti* range must be present. The variations in the magnitudes of the enhancements from molecule to molecule indicated that a different distribution of conformations was required to describe each molecule. The authors (*71*) indicated that more detailed conclusions would require additional studies.

Compound	Irradiate	Observe	Percentage of enhancement	Reference
100		in CDCl$_3$		72
	high field CH$_3$	H–10a	18	
	low field CH$_3$	H–10a	0	
	–OH	H–2	13	
	–OH	H–4	0	
	H–10	–OH	0	
		in C$_5$D$_5$N		
	high field CH$_3$	H–10a	17	
	low field CH$_3$	H–10a	0	
	–OH	H–2	9	
	–OH	H–4	0	
1-Δ^9-Tetrahydro-cannabinol				
101		in CDCl$_3$		
	–OH	H–2	13	
	–OH	H–4	0	
	H–10α (eq)	–OH	0	
		in C$_5$D$_5$N		
	high field CH$_3$	H–10a	25	
	low field CH$_3$	H–10a	0	
	–OH	H–2	13	
	–OH	H–4	0	
	H–10α	–OH	0	
Hexahydrocannabinol				
102		in CDCl$_3$		
	–OH	H–2	13	
	H–10α	–OH	0	
		in C$_5$D$_5$N		
	high field CH$_3$	H–10a	17	
	low field CH$_3$	H–10a	0	
	–OH	H–2	12	
Δ^8-Tetrahydrocannabinol				

Compound	Irradiate	Observe	Percentage of enhancement	Reference
103		in CDCl$_3$		
	–OH	H–2	12	
	–OH	H–4	0	
		in C$_5$D$_5$N		
	–OH	H–2	12	
	–OH	H–4	0	

10[6a(10a)]-Tetrahydrocannabinol

Experimental: The NOE studies were performed at 100 MHz on nitrogen sparged solutions containing approximately 8% (w/v) of the compound in either CDCl$_3$ or C$_5$D$_5$N, as indicated in the tables. TMS was added for field locking. Each resonance was integrated at least ten times with and without the saturating field present, and the NOE's were then calculated from the average areas.

Conclusions: In **100–102**, saturation of the high field methyl resonance resulted in a substantial enhancement of the H–10a resonance. Saturation of the low field methyl had no effect on the H–10a resonance. Thus the high field methyl line was assigned to the 6α–CH$_3$ and the low field methyl resonance to 6β–CH$_3$. In addition, of the two conformations considered possible for the B ring, only the one drawn allows sufficiently close approach of the 6α–CH$_3$ to H–10a to account for the observed NOE. Therefore this conformation was assigned to the B ring.

The enhancements observed between H–2 and the –OH proton in **100–103** allowed the H–2 resonances to be assigned and further indicated that the predominant conformation is that with H–2 *syn* to the –OH group.

D. Miscellaneous Applications of the NOE

The earliest applications of the nuclear Overhauser effect were in studies of nuclear spin relaxation. The work of Solomon (*73a*) and Solomon and Bloembergen (*73b*) on liquid HF, which has already been discussed in Section 7,C, falls into this category. Hubbard (*74*) has shown that NOE data and spin relaxation times measured at a single temperature can be used to determine the relative contributions of spin–rotation and dipole–dipole interactions to the relaxation of a spin-$\frac{1}{2}$ nucleus. Chaffin and Hubbard (*75*) applied this technique to study the relaxation of both fluorines and protons in neat CHF_3 from 113 to 173°K. They found that the intermolecular dipole–dipole mechanism was more important than the intramolecular dipole–dipole mechanism and that the spin–rotation mechanism is important for the fluorine but not for the protons.

Mackor and Maclean (*76*) made a similar application of fluorine–proton Overhauser effects in their study of spin relaxation in $CHFCl_2$. They found that at temperatures just above the melting point both protons and fluorine spins relaxed primarily through interactions between unlike spins, with the intramolecular dipole–dipole mechanism being dominant. At 293°K the protons appeared to relax primarily through intermolecular dipole–dipole interactions while spin–rotation becomes the major mechanism for the fluorine nuclei. Brown *et al.* (*77*) had arrived at similar conclusions in an earlier study of this molecule.

Kuhlmann *et al.* (*78*) have reported a study of the ^{13}C relaxation times and $^{13}C–\{H\}$ Overhauser effects in adamantane (**104**). In this paper they show that the $^{13}C–\{H\}$ NOE of a single carbon relaxing primarily through its dipole–dipole interaction with a group of equivalent protons should be independent of the number of protons in the group (see the discussion in Section 3,B). Under the same conditions, the ^{13}C relaxation time can be shown to be inversely proportional to the number of protons in the group. Adamantane was chosen to test the theory because (1) it is rigid and highly symmetrical so interpretation of the results would not be complicated by the possibility of asymmetric tumbling or internal motions and (2) it contains two types of magnetically nonequivalent carbons, one bonded to a single proton and the other to two equivalent protons. Very good agreement was found between experiment and theory. The Overhauser enhancements of the two carbon resonances were equal to within experimental error when the proton resonances were saturated, and the ratio T_{1CH}/T_{1CH_2} was found to be 1.80 ± 0.25. The ratio T_{1CH}/T_{1CH_2} would be 2.0 if the ^{13}C nuclei relaxed entirely through their dipolar interaction with the directly bonded protons, but taking next nearest protons into account leads to a theoretical value of 1.82—in excellent agreement with the observed ratio.

TABLE 8-7

NUCLEAR OVERHAUSER EFFECTS IN METHYL
METHACRYLATE (**105**): 10% SOLUTION IN CS_2[a]

	Irradiate			
Observe	H–C	H–O	H–H	H–L
H–C	—	0	1	0
H–O	0	—	0	0
H–H	9	0	—	42
H–L	1	0	48	—

[a] Data from Fukumi *et al.* (*79*). The data are accurate to ±1%.

Fukumi *et al.* (*79*) also used a combination of NOE and relaxation measurements in a study of methyl methacrylate (**105**). The experiments were performed at 100 MHz and 28°C on both a neat sample and a 10% solution of **105** in CS_2. The observed enhancements are recorded in Tables 8-7 and 8-8. The NOE values confirmed previous assignments of the spectrum.

Fukumi *et al.* also found that in the 10% solution in CS_2, relaxation was primarily due to intramolecular dipole–dipole interactions. This was indicated by the fact that further dilution of the sample with CS_2 caused very little change in either the NOE's or the T_1 values. The large values for $f_H(L)$ and $f_L(H)$ in this solution plus the similarity of the T_1's of L and H indicated that these spins relax almost entirely through their mutual dipole–dipole interaction; the small differences could be attributed to the additional relaxation of H through interaction with the C-methyl group. Fukumi *et al.* also noted that

$$T_1(O) = 16.0\,\text{sec} > T_1(C) = 9.3\,\text{sec}$$

this inequality presumably being the result of rather rapid rotation of the O-methyl about the carbon–oxygen bond.

TABLE 8-8

NUCLEAR OVERHAUSER EFFECTS IN METHYL
METHACRYLATE (**105**): NEAT SAMPLE[a]

	Irradiate			
Observe	H–C	H–O	H–H	H–L
H–C	—	4	2	1
H–O	6	—	2	2
H–H	9	5	—	28
H–L	6	5	29	—

[a] Data from Fukumi *et al.* (*79*). The data are accurate to ±1%.

In the neat sample of **105**, all relaxation times were less than in the CS_2 solution, indicating a significant contribution from intermolecular relaxation. The values of $f_L(H)$ and $f_H(L)$ also decrease, reflecting the increased importance of the intermolecular mechanism. However, other enhancements [e.g., $f_O(C)$ and $f_L(C)$] increase. These increases must be interpreted as intermolecular Overhauser effects. This is one of the few confirmed cases of an intermolecular NOE appearing in the literature.

The first observation of an intermolecular NOE was made by Kaiser (*80*) on a solution containing chloroform, cyclohexane, and TMS in the approximate mole ratios 1:4.5:0.5. Saturation of the cyclohexane resonance in this sample led to a 31% enhancement of the signal from the chloroform proton.

Chan and Kreishman (*81*) have made an interesting application of the intermolecular NOE in a study of purine (**106**) intercalation in 3′,5′-polyuridylic acid (**107**). The experiments were done at 220 MHz on a 0.4 M purine solution containing an amount of polyuridylic acid equivalent to 0.10 M uridine. The results are summarized in Table 8-9, and suggest that the purines preferentially orient themselves in the complex with either the H–6 or H–8 protons directed at the H–3′, H–5′, and H–5″ protons of the ribose.

TABLE 8-9

INTERMOLECULAR NUCLEAR OVERHAUSER EFFECTS OBSERVED
BETWEEN POLYURIDILIC ACID AND PURINE[a]

Purine proton observed	Ribose Proton Irradiated			
	H–5′ and H–5″	H–2′ and H–4′	H–3′	H–1′
H–6	11 ± 3	0	0	0
H–2	0	0	0	0
H–8	0	0	6 ± 3	0

[a] Data from Chan and Kreishman (*81*).

Another application of the NOE that has been reported in the literature is for improvement of the signal-to-noise ratio in ^{13}C spectra (*82–85*). The procedure usually used is to saturate all proton resonances using a "white noise" decoupler while the carbon spectrum is recorded. If the ^{13}C spins relax primarily through dipolar interactions with neighboring protons, this technique can increase the amplitude of the ^{13}C spectrum by a factor of 2.99 (see Table 3-8). Additional improvement in the sensitivity can, of course, be realized if the ^{13}C spectrum is observed using the Fourier transform technique (*85*). Feeney *et al.* (*84*) have noted that this sensitivity enhancement can also be obtained without loss of the ^{1}H–^{13}C spin–spin splittings if the proton decoupler is turned off just before recording the spectrum; this procedure is most amenable to use with the Fourier transform method.

Anet (*86*) has applied the NOE to assign the lines in the proton NMR spectra of adducts of thallium (III) acetate with norbornene (**108**) and norbornadiene (**109**). Naturally occurring thallium consists of 70.5% ^{205}Tl and 29.5% ^{203}Tl, both of which are spin-$\frac{1}{2}$ nuclei. However, the ^{1}H–^{205}Tl coupling constants are only about 1% higher than the corresponding ^{1}H–^{203}Tl coupling constants (*87*), so for purposes of interpreting the NMR spectrum, the thallium–proton splittings can be handled as if only a single thallium isotope were present.

The ^{1}H NMR spectrum of **108** showed that at least five protons had very large spin–spin couplings with thallium, but the pair of lines corresponding to a given proton could not be picked out directly due to overlap of lines and the absence of fine structure in the spectrum. However, Anet noted that saturation of one line of a proton doublet could cause partial saturation of the other half of the doublet providing the relaxation time of the thallium was shorter than that of the protons. (Strictly speaking, this is a generalized Overhauser effect, cf. page 8.) The reason for this is that one line of the doublet corresponds to thallium nuclei in the spin state $+\frac{1}{2}$ and the other to thallium nuclei in the spin state $-\frac{1}{2}$. Relaxation processes acting in the solution result in the thallium nuclei continually changing their spin states so that protons corresponding to one line of the doublet become transferred to the other line. The transfer of protons between halves of the doublet results in the transfer of magnetization between sites as well (see Section 7,B), so that saturation of one

TABLE 8-10

PROTON–FLUORINE NUCLEAR OVERHAUSER EFFECTS[a,b]

Compound	Percentage of enhancement	
	^{19}F–{^{1}H}	*ortho* ^{1}H–{^{19}F}
2,4-Dinitrofluorobenzene	22	10
p-Fluoronitrobenzene	45	11
1-Fluoronapthalene	50 (48[c])	—
o-Fluorotoluene	28	—
	18 {H}	—
	11 {CH$_3$}	—

[a] Data from Bell and Saunders (*88*).

[b] Experiments were performed on degassed solutions containing 20% of the compound, 20% CCl$_3$CF$_3$ for field-frequency locking, 25% (CD$_3$)$_2$SO, and 25% CDCl$_3$.

[c] The (CD$_3$)$_2$SO was replaced by benzene in this sample.

line leads to partial saturation of the other line. Making use of this effect, Anet was able to completely assign the spectra for both **108** and **109**.

A few other reports should also be mentioned. Bell and Saunders (*88*) have measured ^1H–^{19}F NOE's in several fluoro-aromatic compounds. Their results are summarized in Table 8-10. Natusch and Richards (*89*) have reported on experiments in which both an electron resonance and a nuclear resonance were saturated while a second nuclear resonance was observed. These experiments were done on a 1:1 mixture of benzene and hexafluorobenzene containing approximately 5×10^{-4} M tri-*t*-butyl-phenoxyl radicals. Finally, the report by Bell and Saunders (*90*) on the correlation between internuclear distances and the intramolecular nuclear Overhauser effect has already been discussed on page 51.

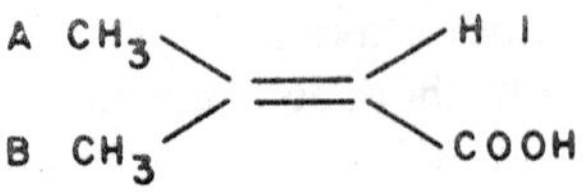

A CH3
B CH3
H
COOH
I
1

CH3
H3C
CH3
3
4
2
5
10
CH3
6

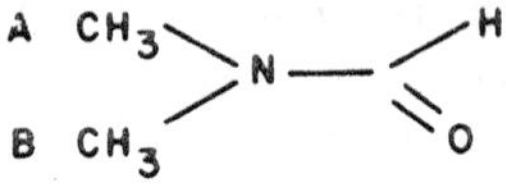

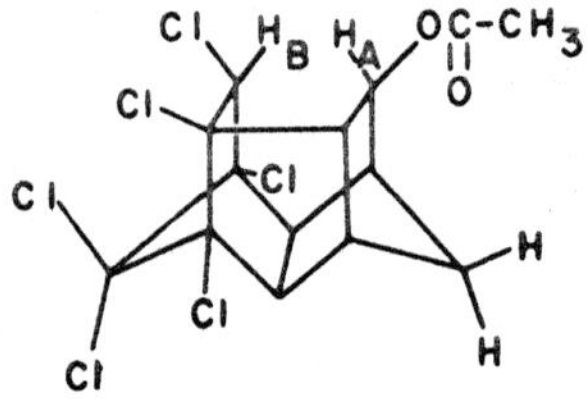

A CH3
B CH3
N
H
O
2

H
4
1
H
CH3
CH3
H anti
H syn
7

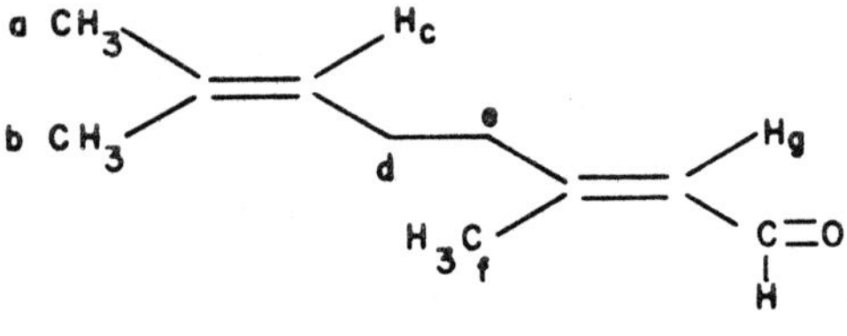

Cl
Cl
Cl
Cl
Cl
Cl
HB
B
HA
A
OC-CH3
O
H
H
3

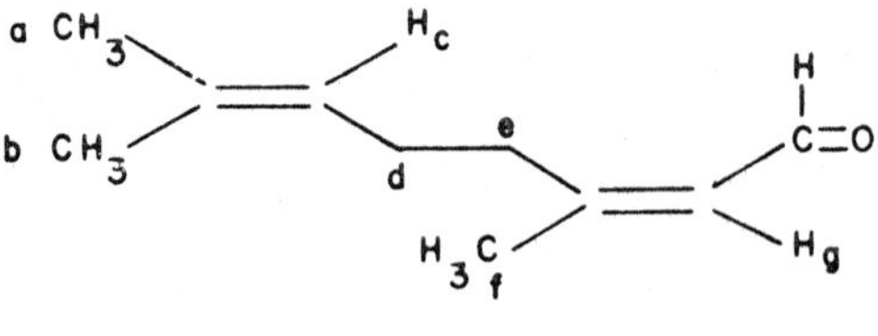

CH3
O
H3C
N
N
N
N
O
CH3
H
8

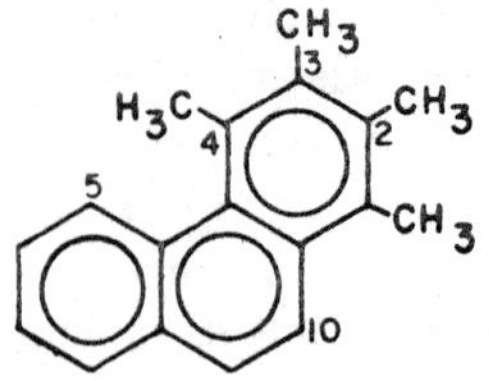

a CH3
b CH3
Hc
d
e
H3C
f
Hg
C=O
H
4

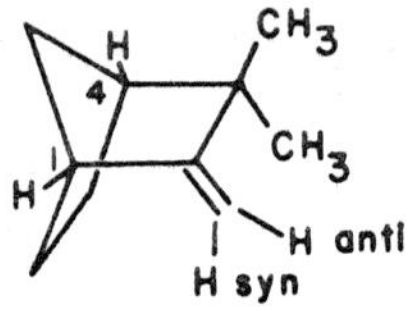

a CH3
b CH3
Hc
d
e
H3C
f
H
C=O
Hg
5

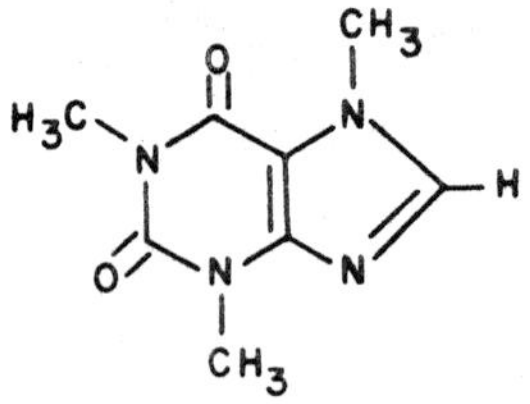

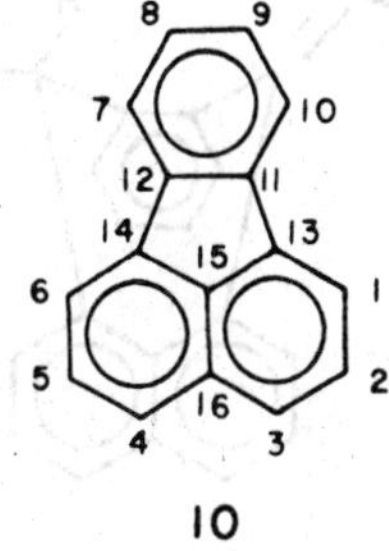

9

10

11

12

13

14

15

16

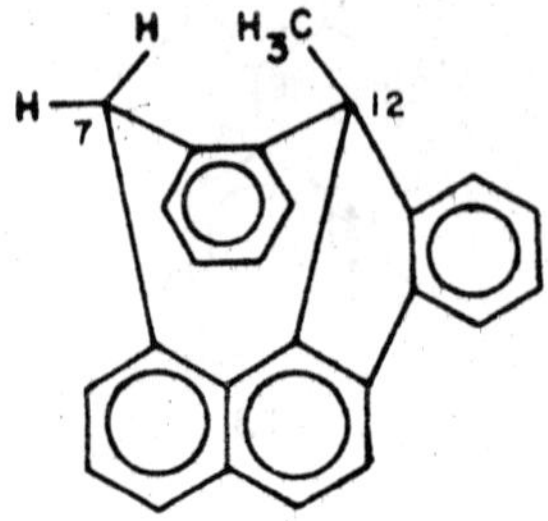

H
H₃C
12
7
17

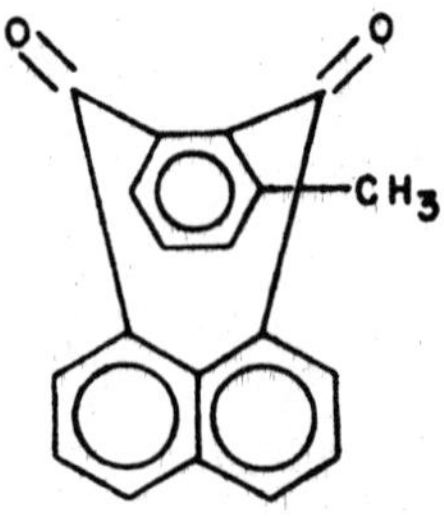

O
O
CH₃
22

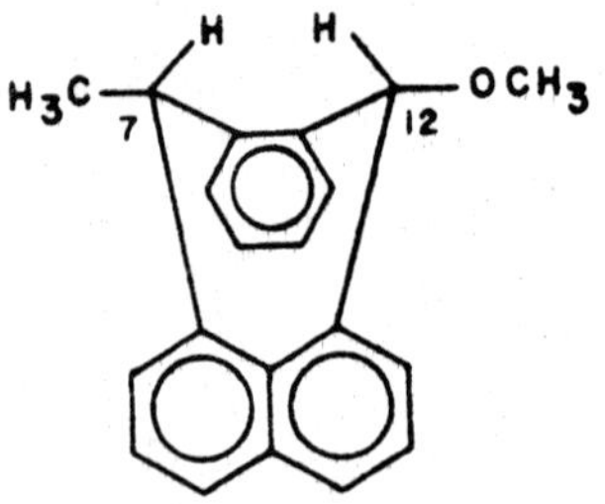

H₃C
H
OCH₃
7
12
18

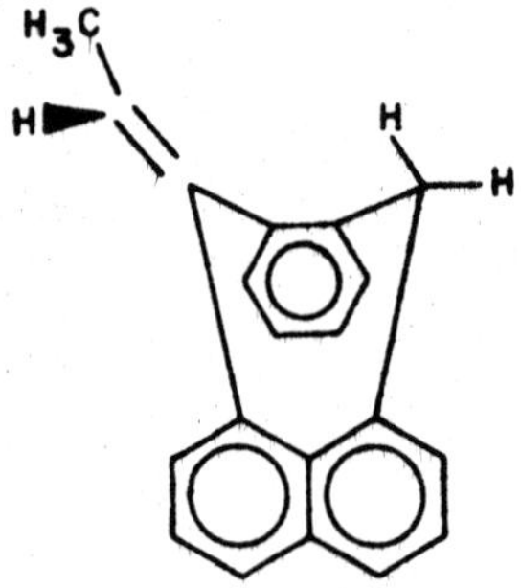

H₃C
H
H
H
23

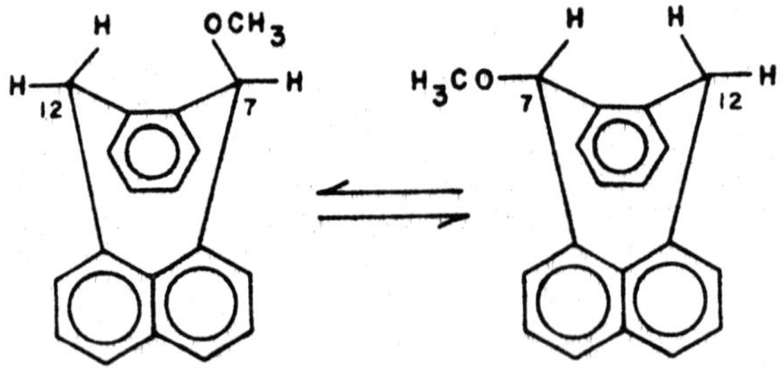

H
OCH₃
H
H
H
H₃CO
H
12
7
7
12
19

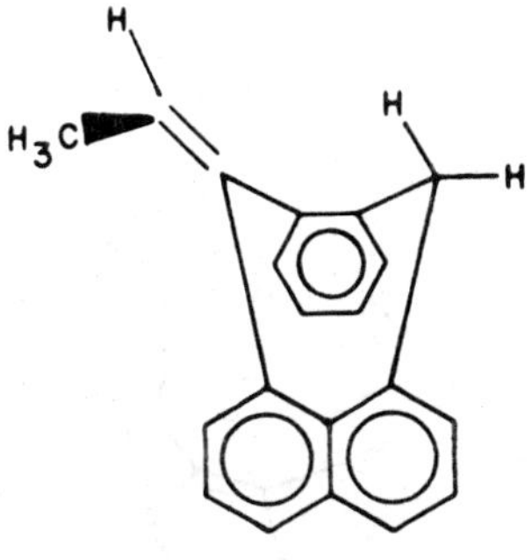

24

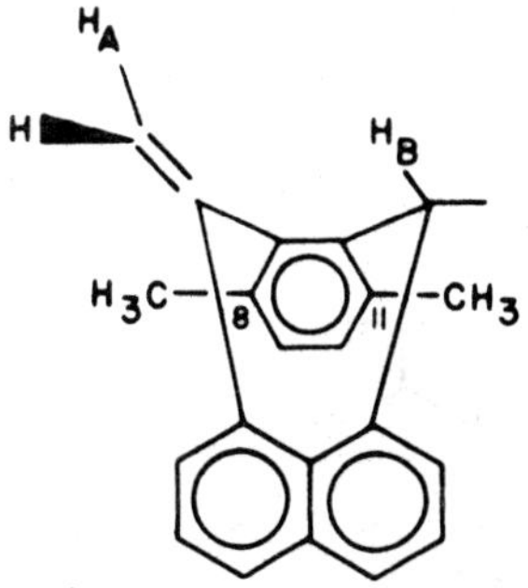

27

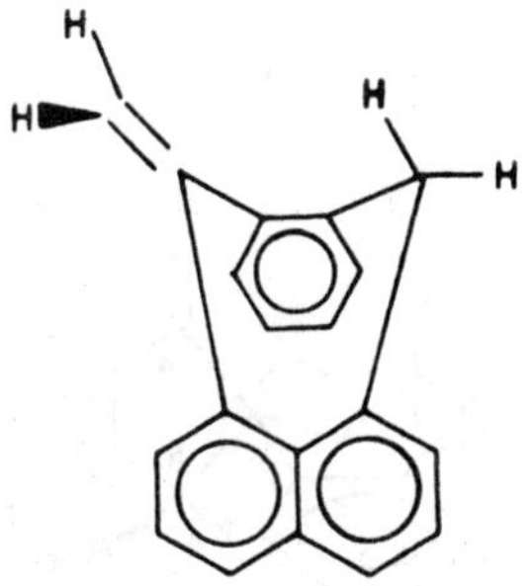

25

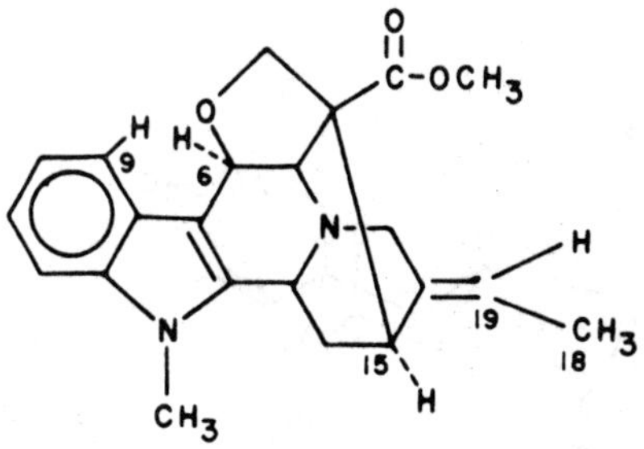

28

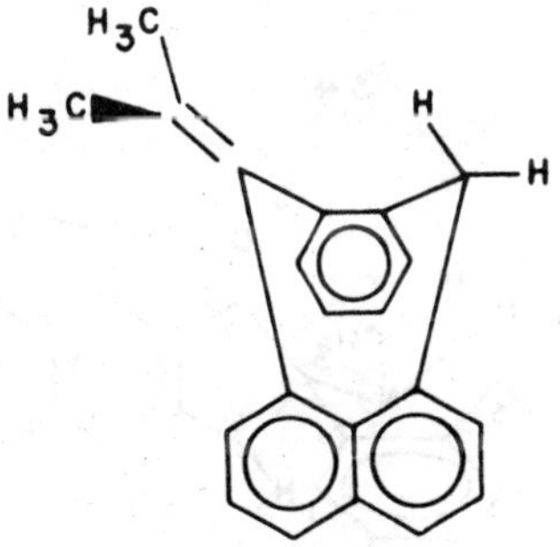

26

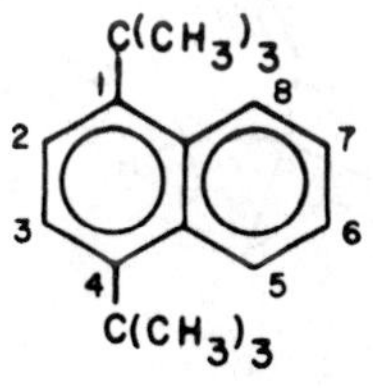

29

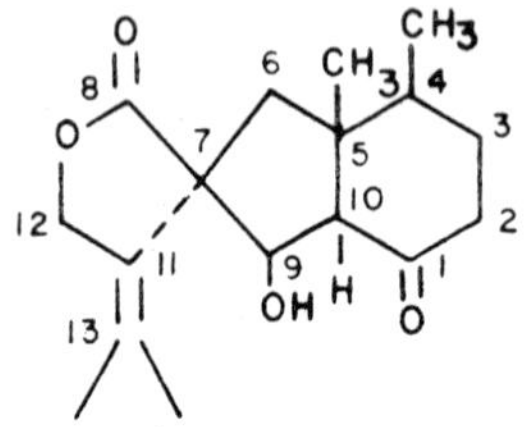

30

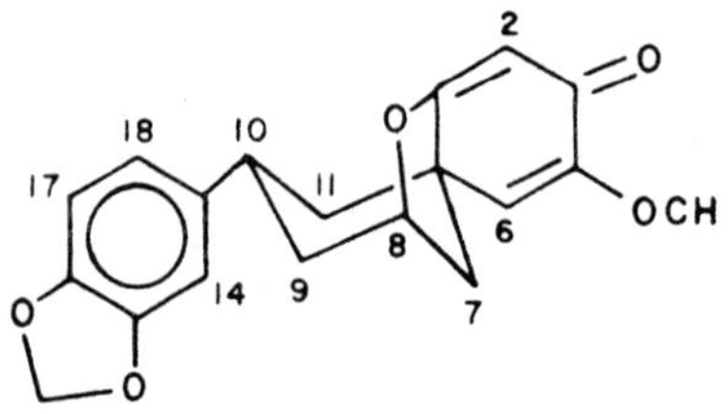

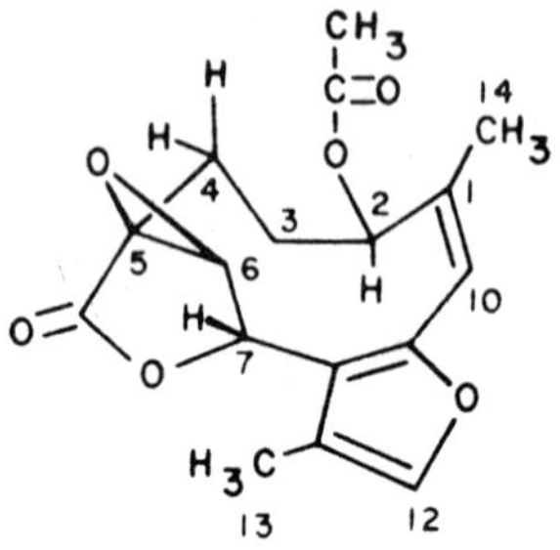

33

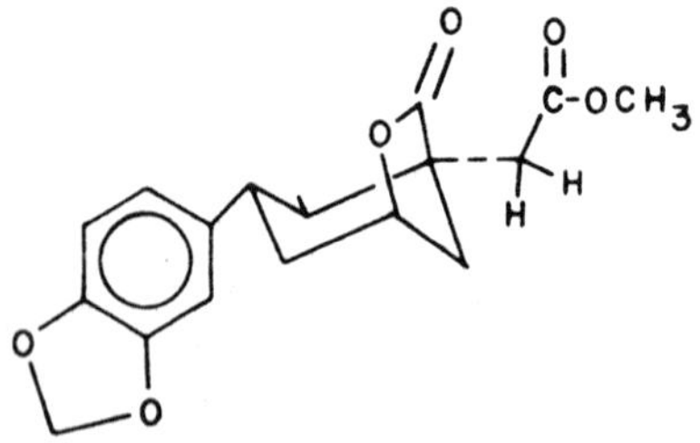

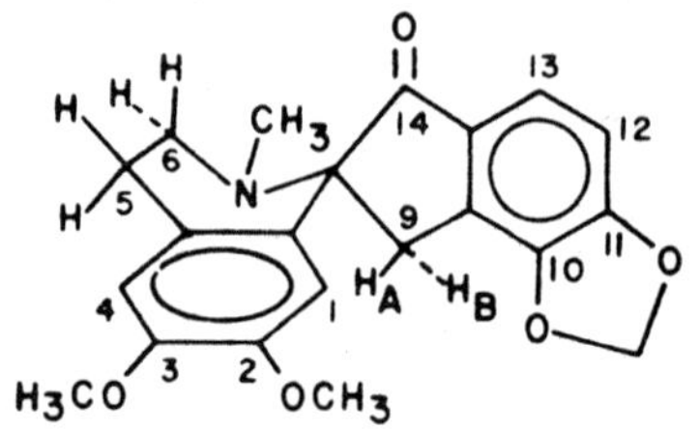

31

34

32

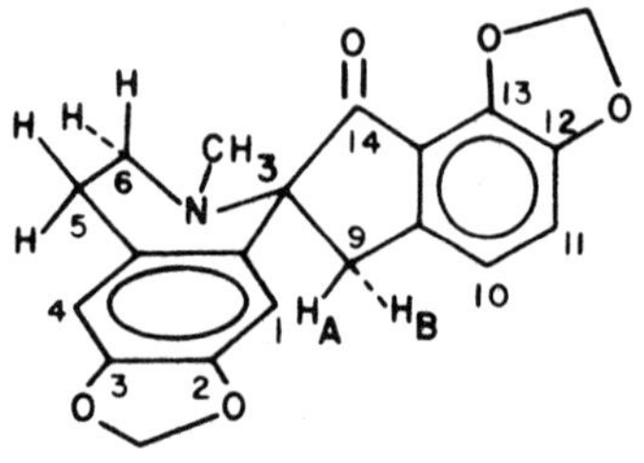

35

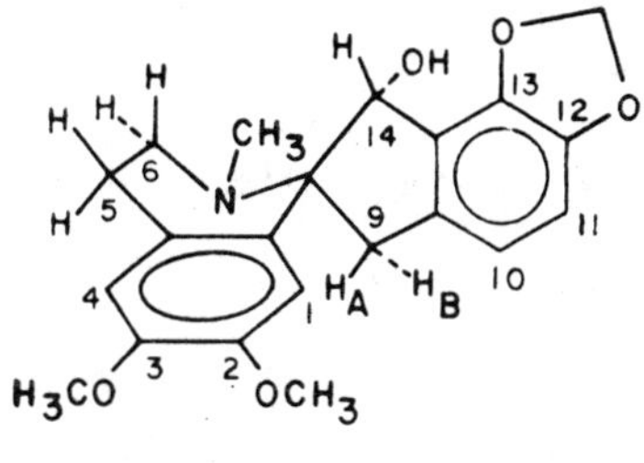

36

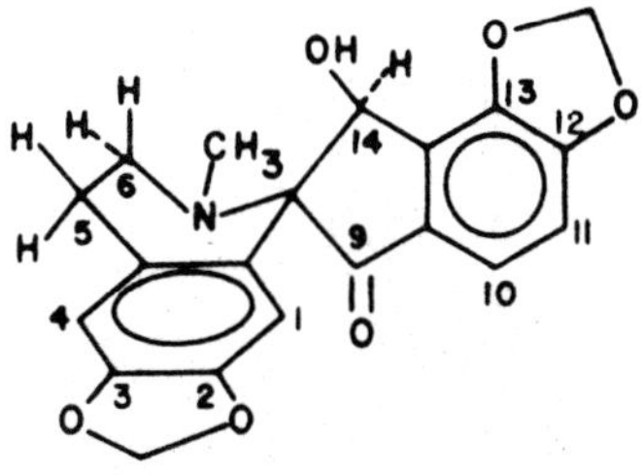

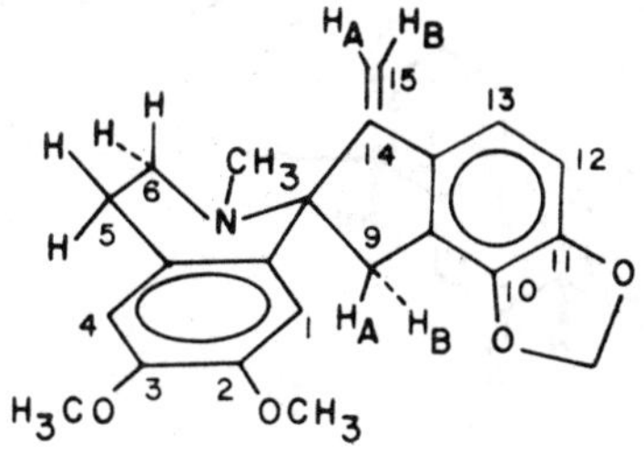

37

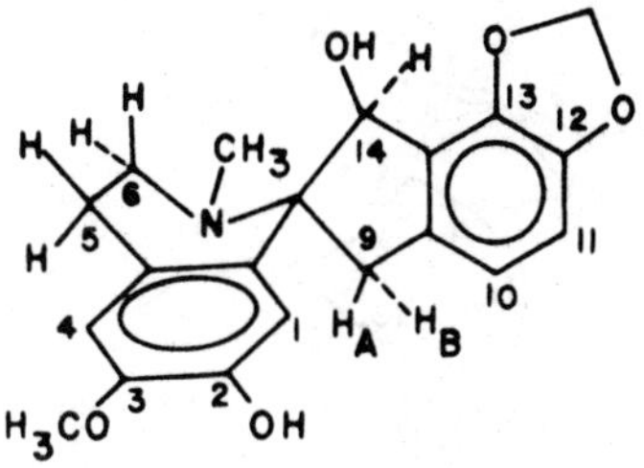

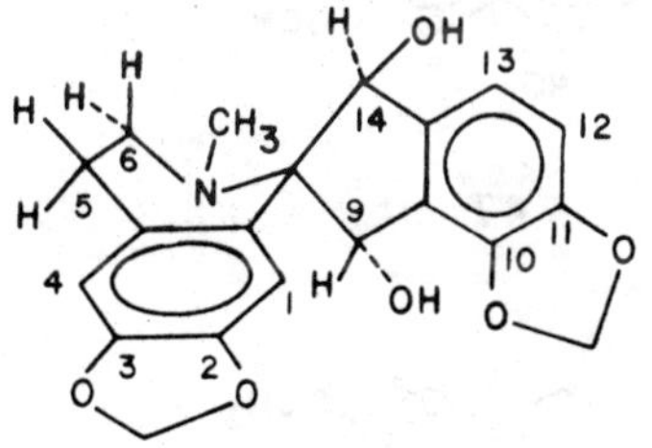

38

39

40

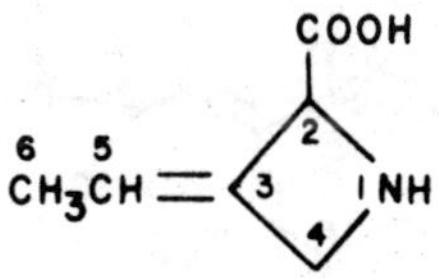

41

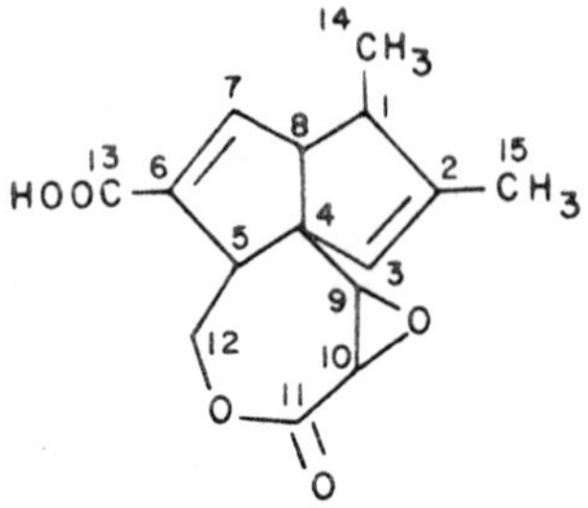

42

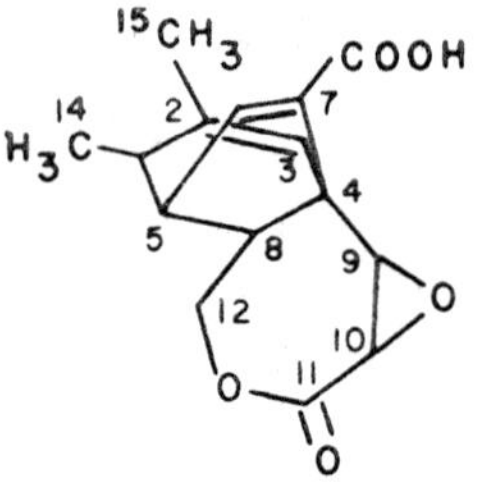

43

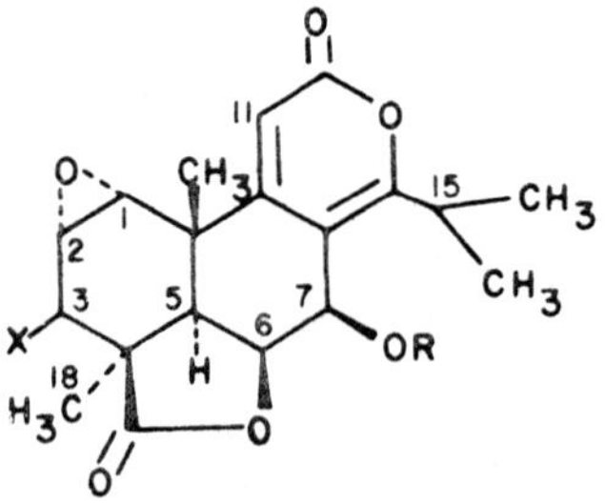

44

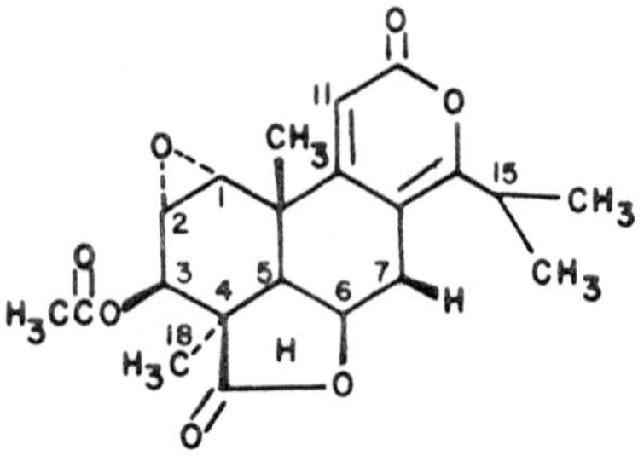

45

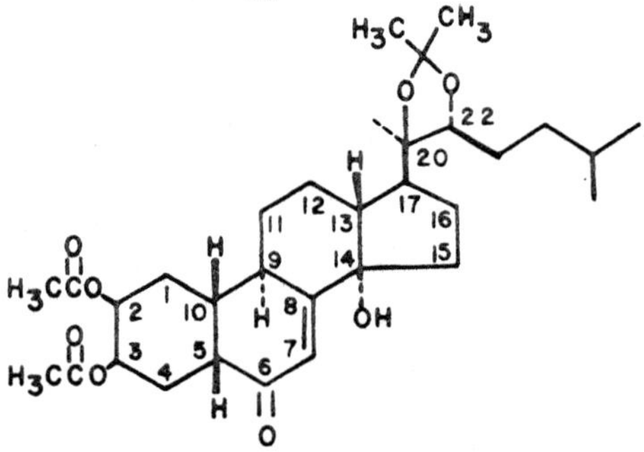

46

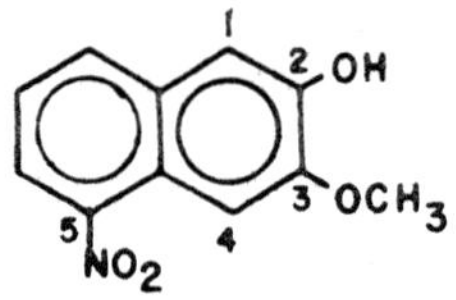

47

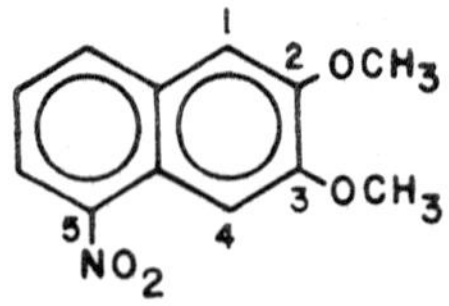

48

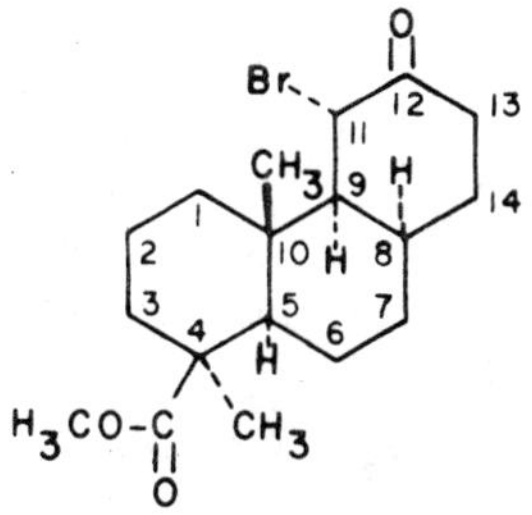

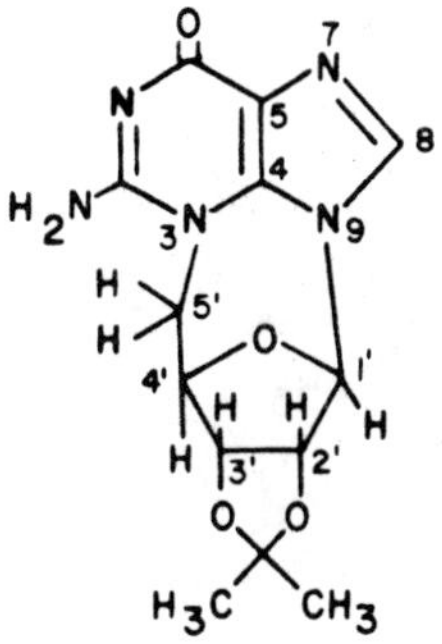

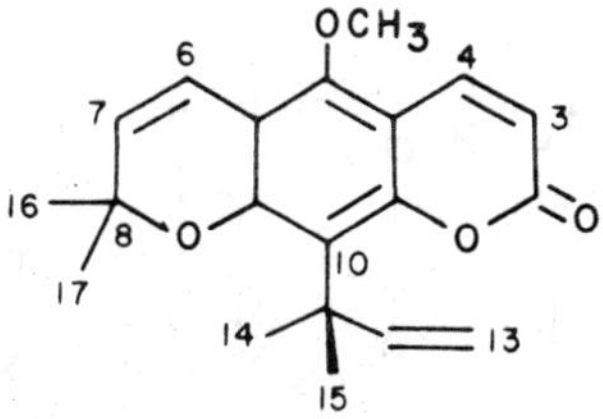

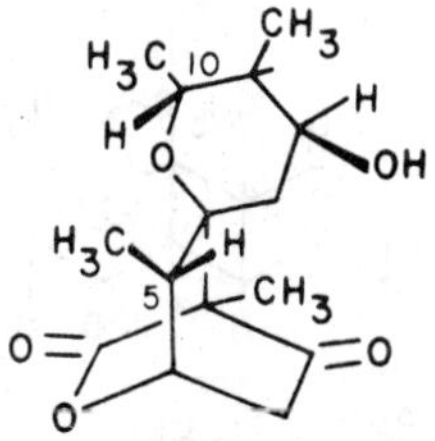

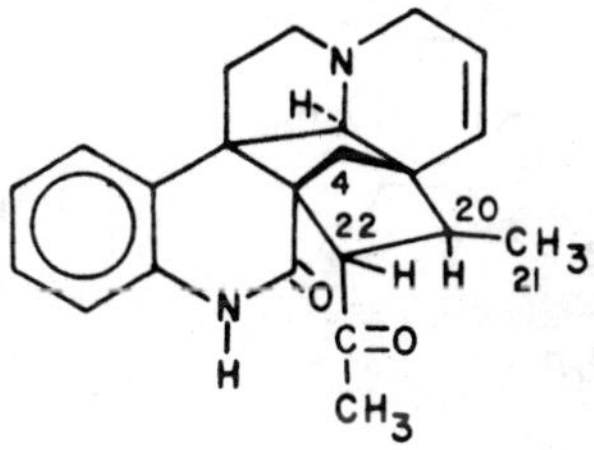

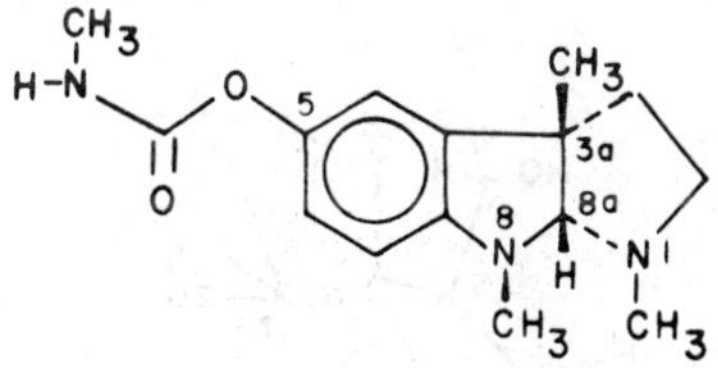

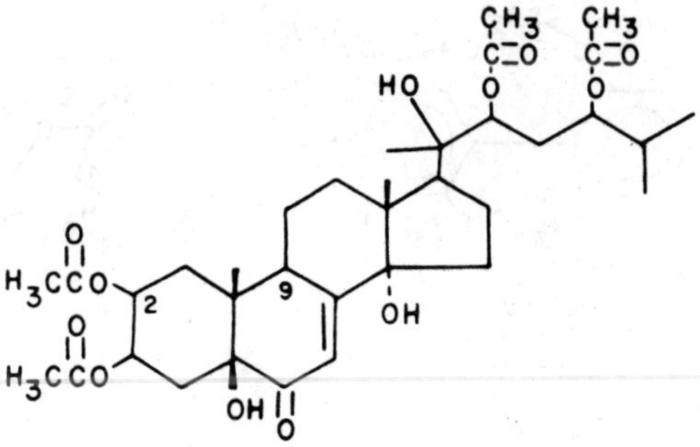

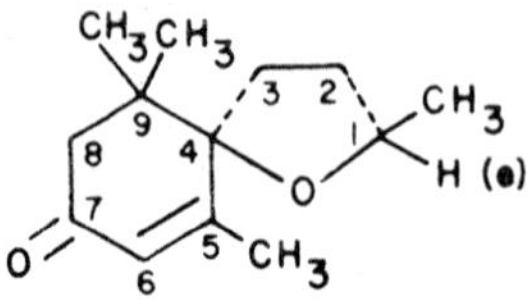

56
57 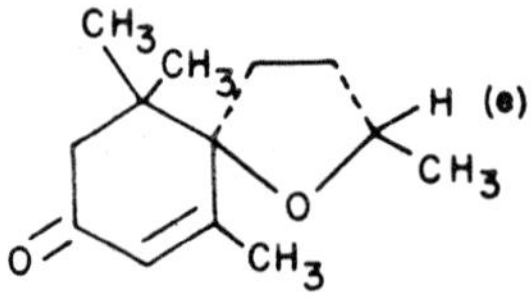
58 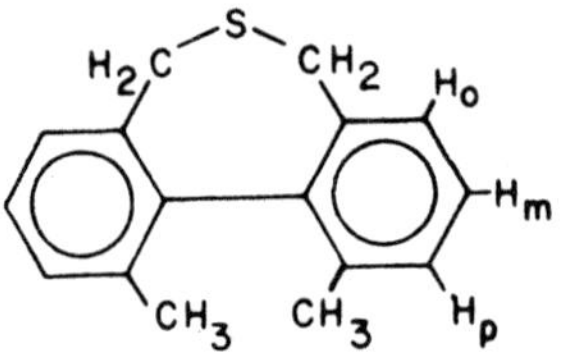
59 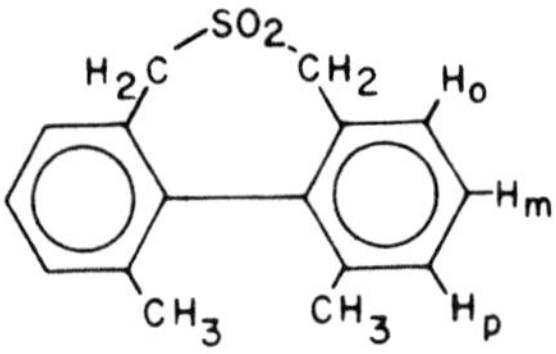
60 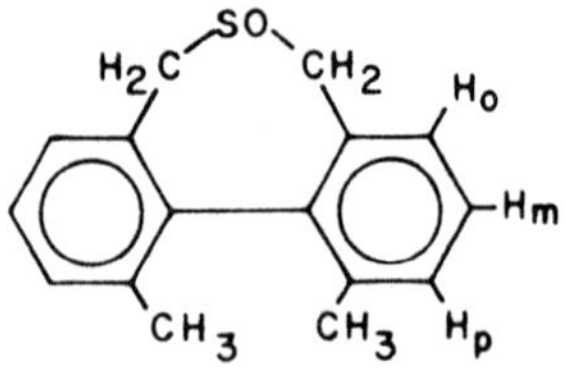
61 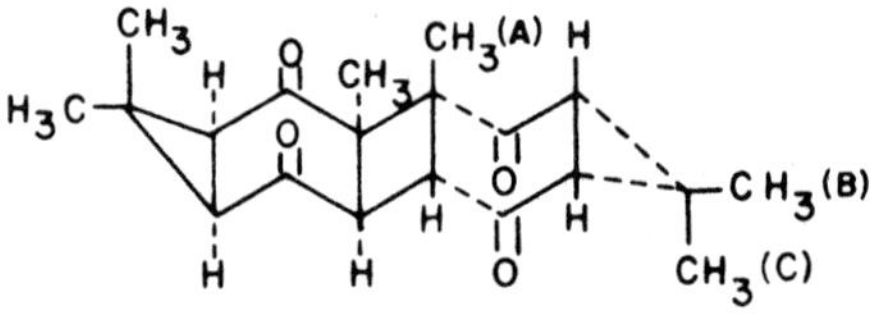
62 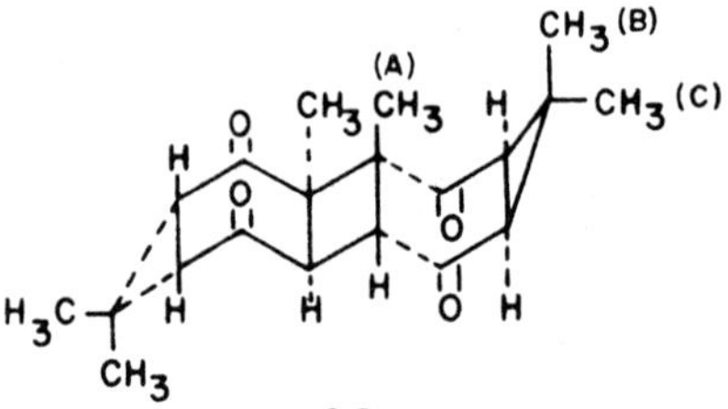
63

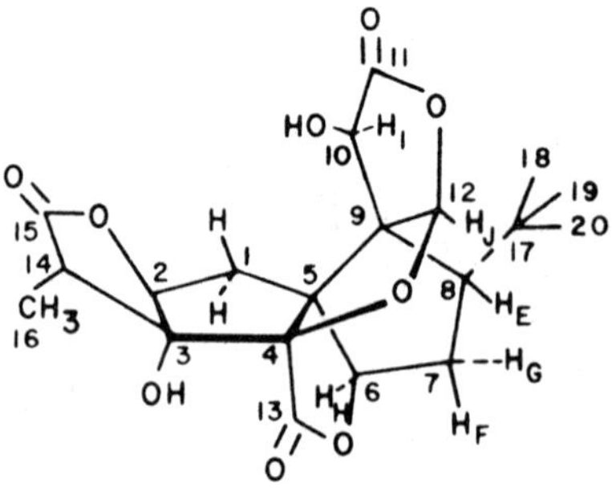

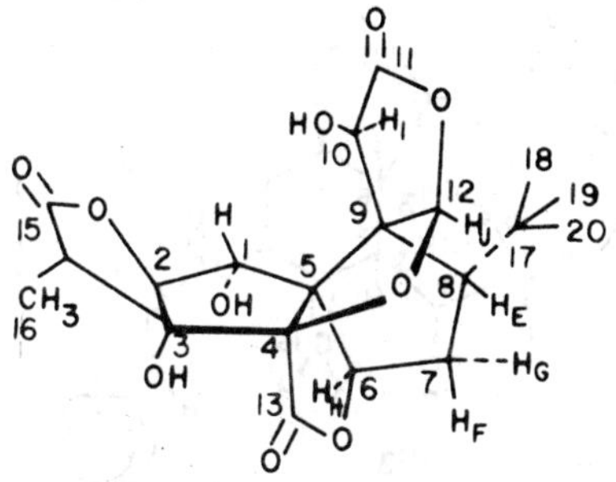

64

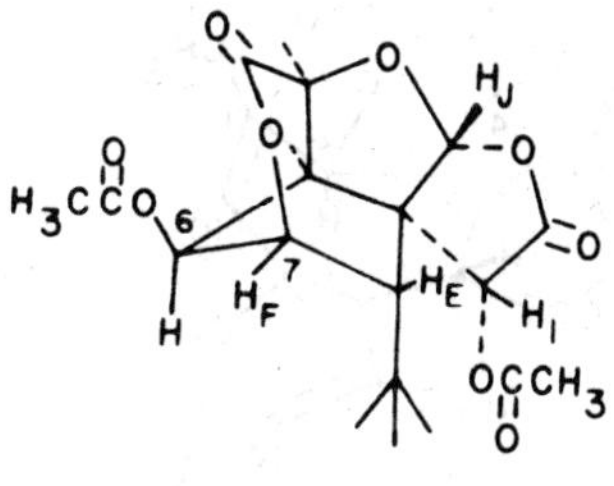

67

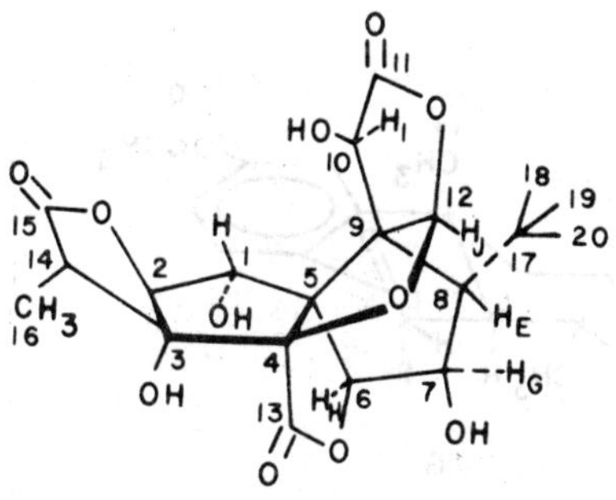

65

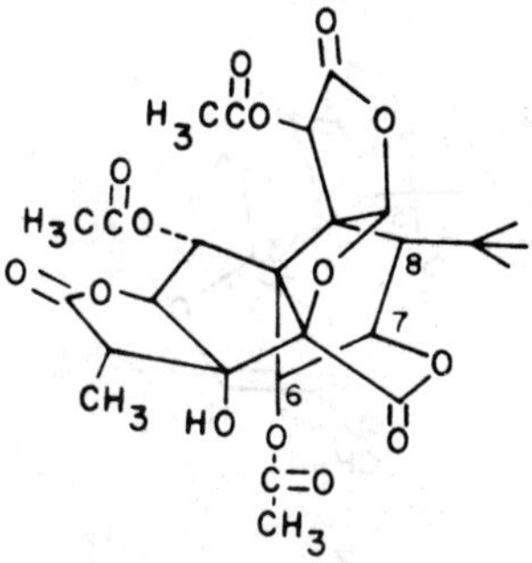

68

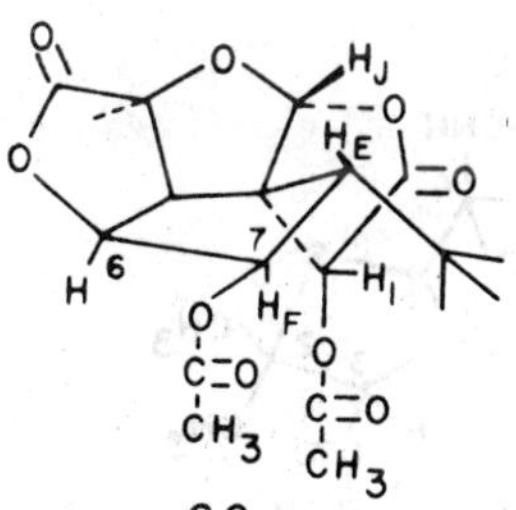

66

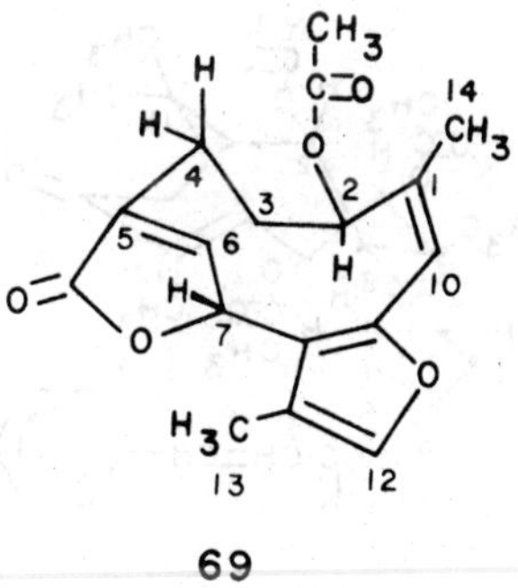

69

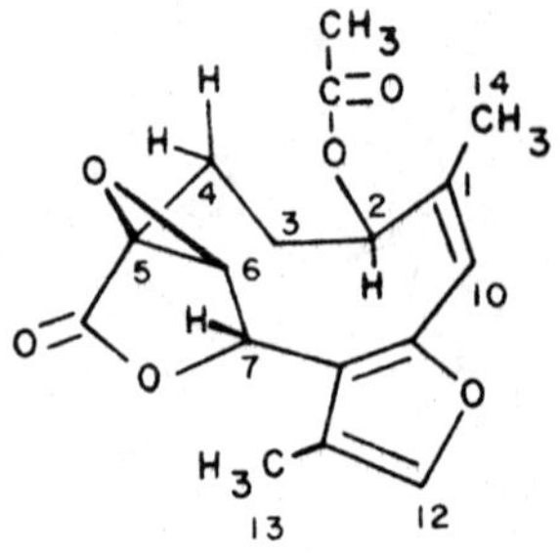

70
73-75 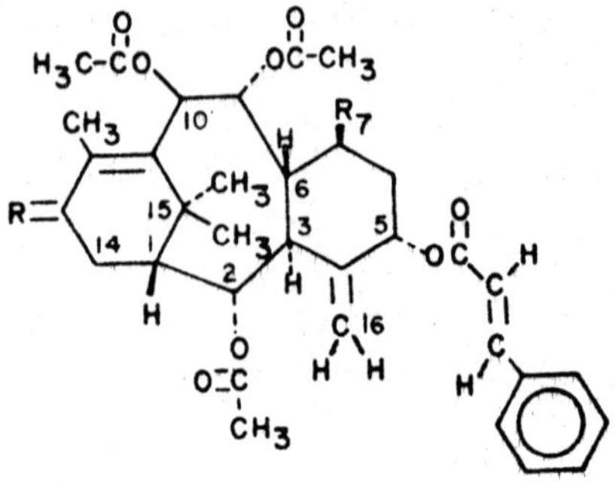
71 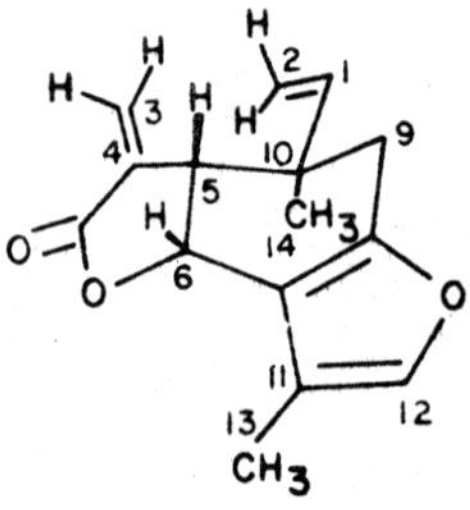
76 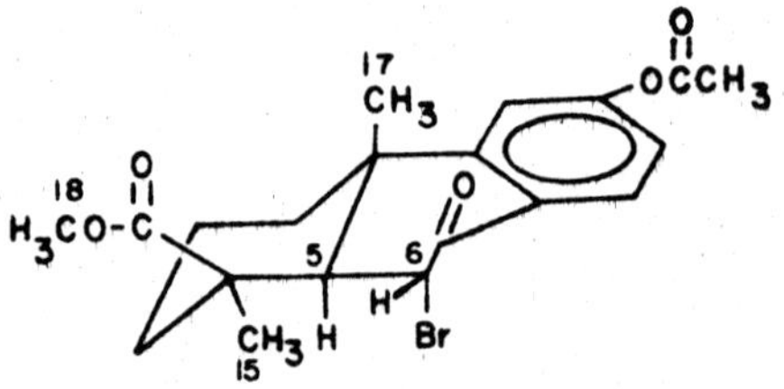
72 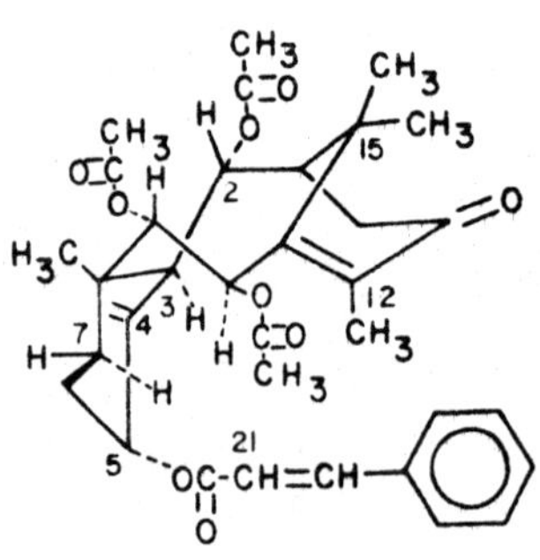
77

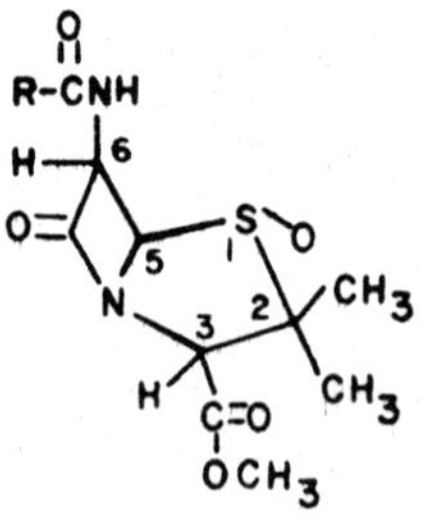

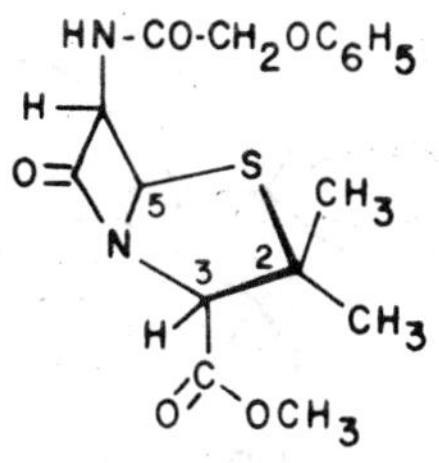

78

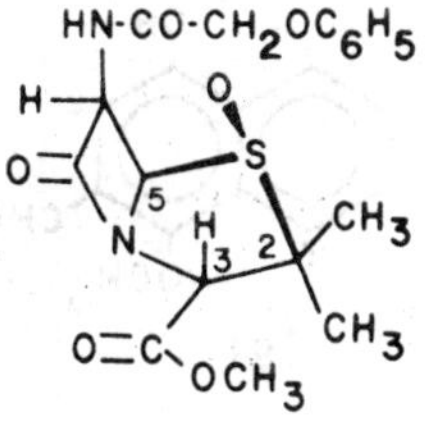

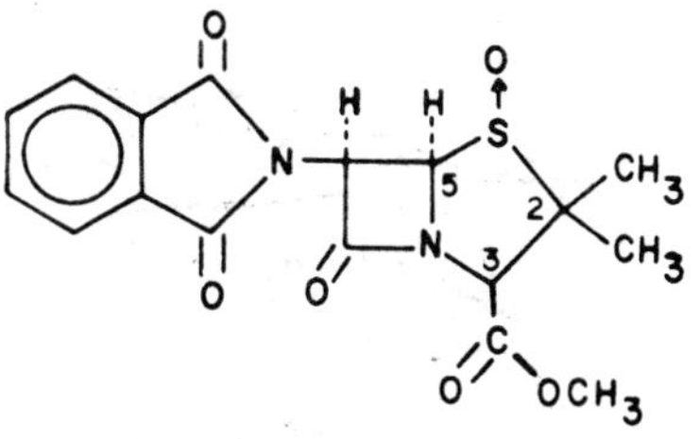

82

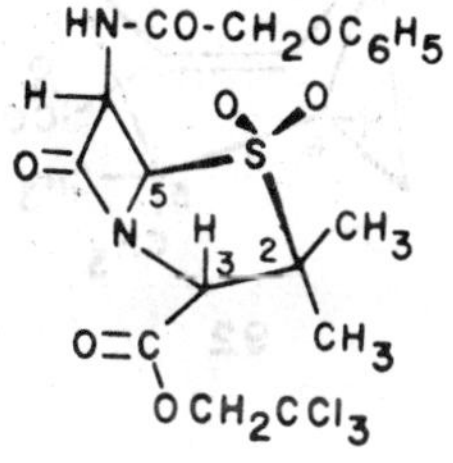

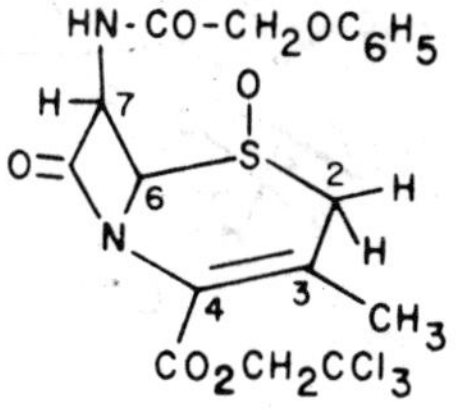

79

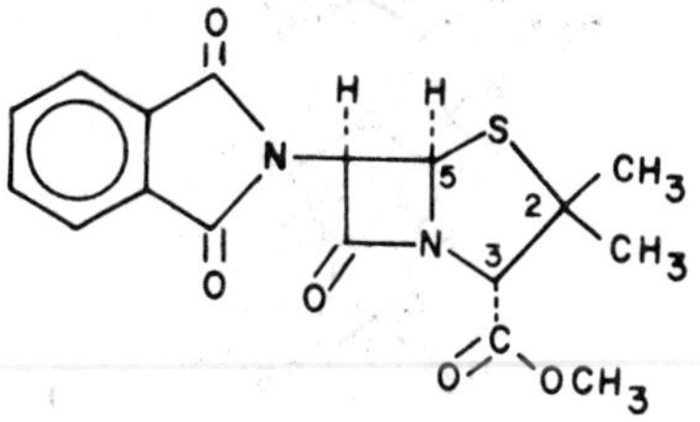

83

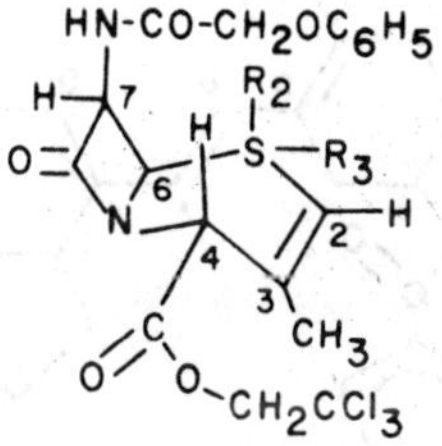

80

84

81

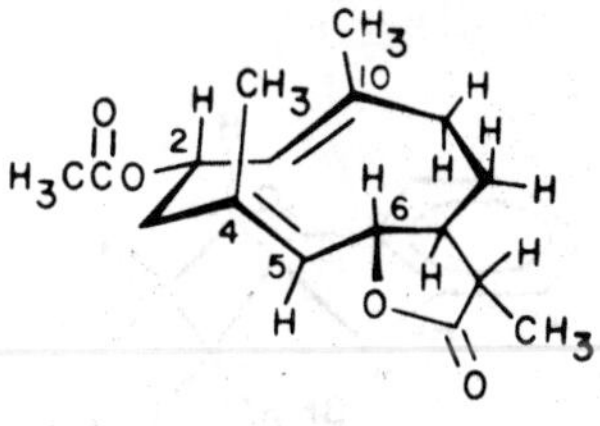

85

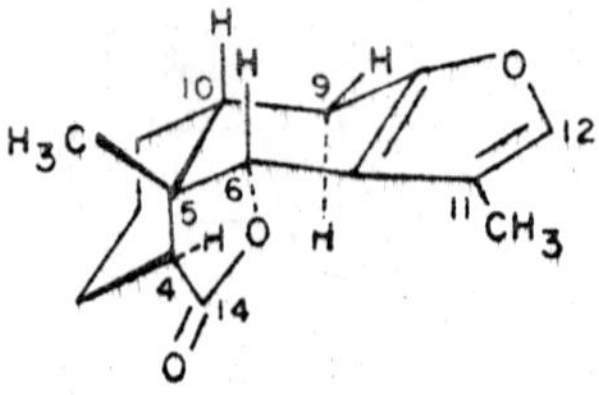

86

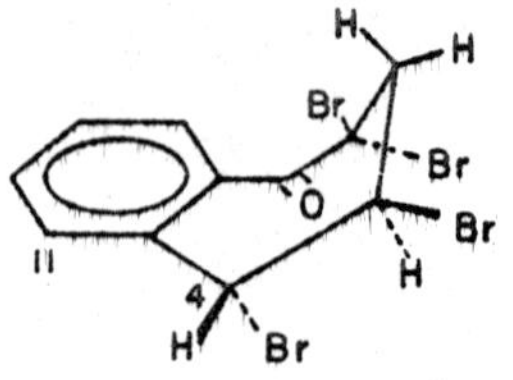

90

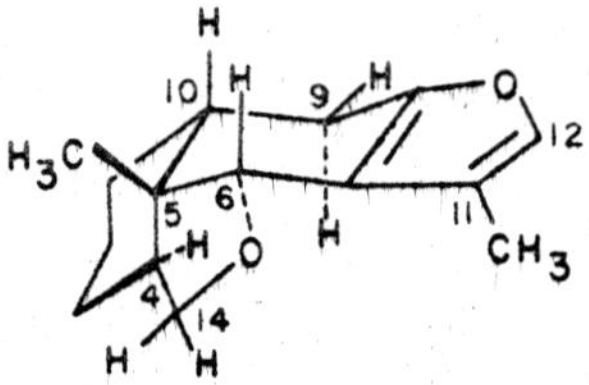

87

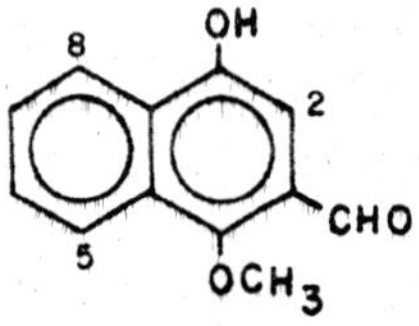

91

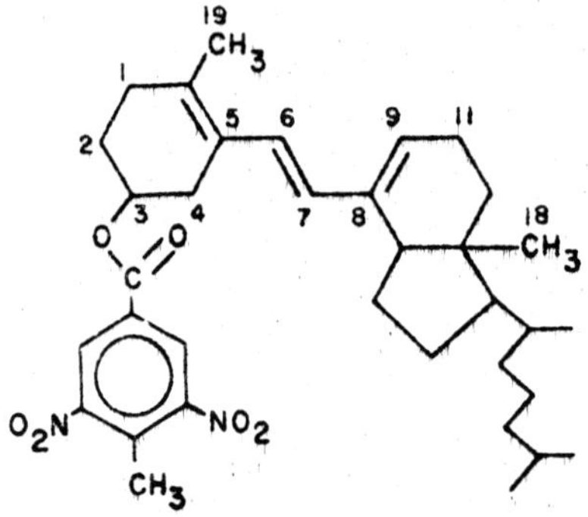

88

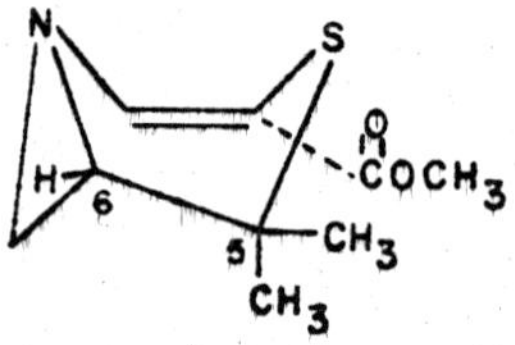

92

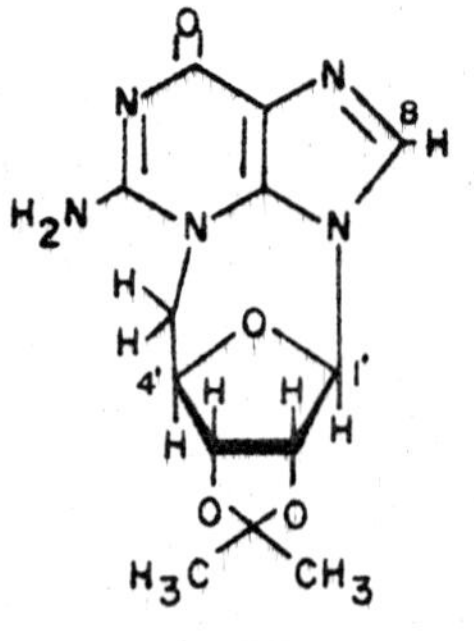

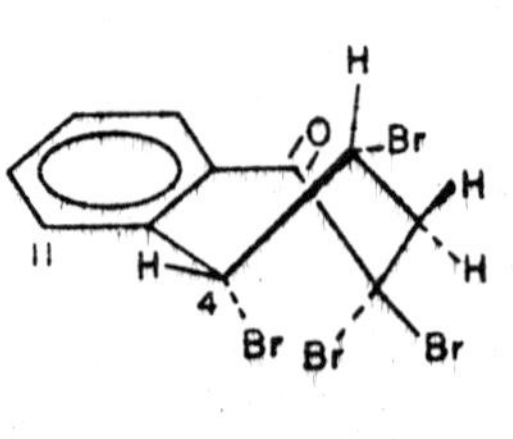

89

93

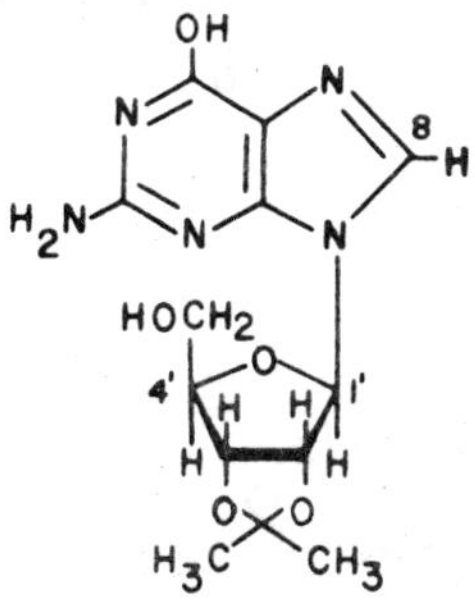

94
95
96 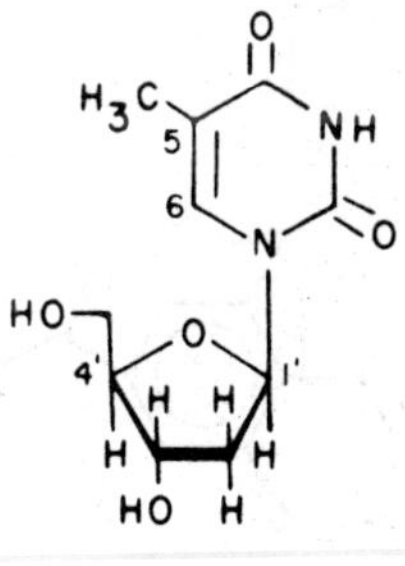
97
98
99

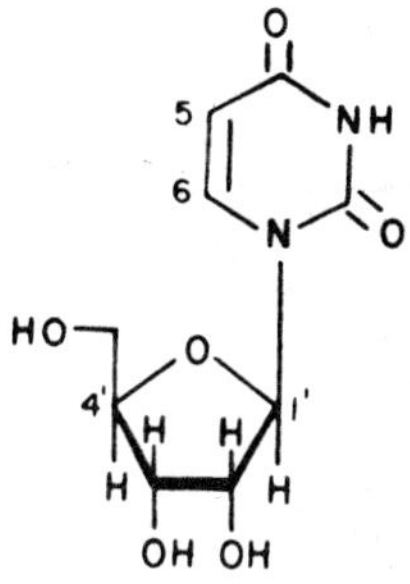

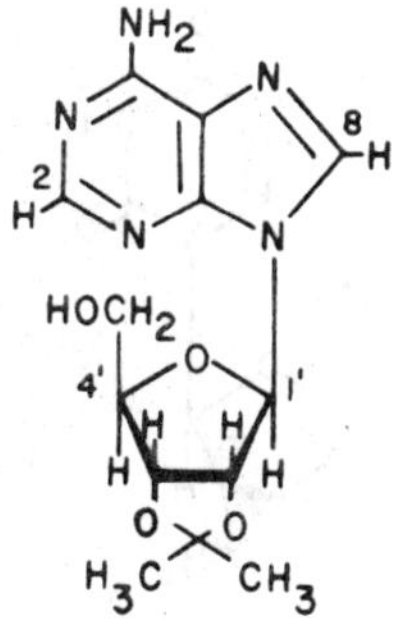

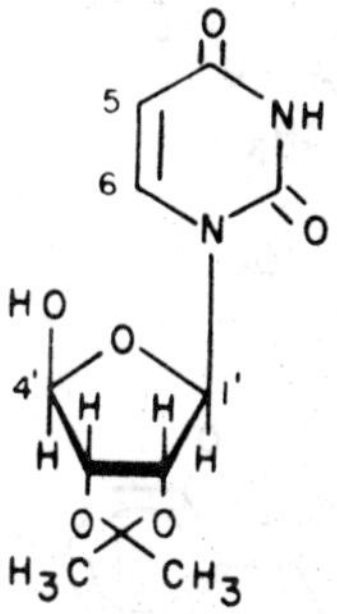

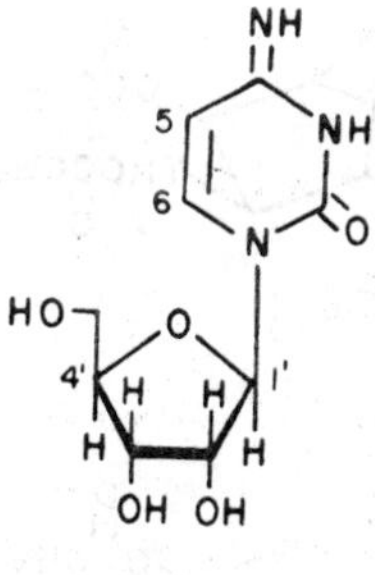

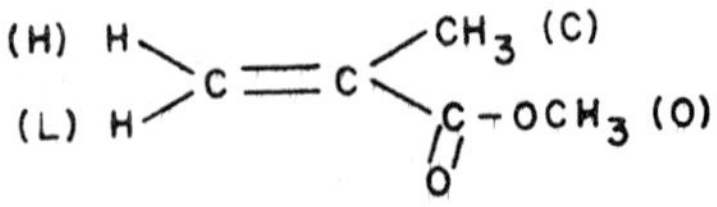

100

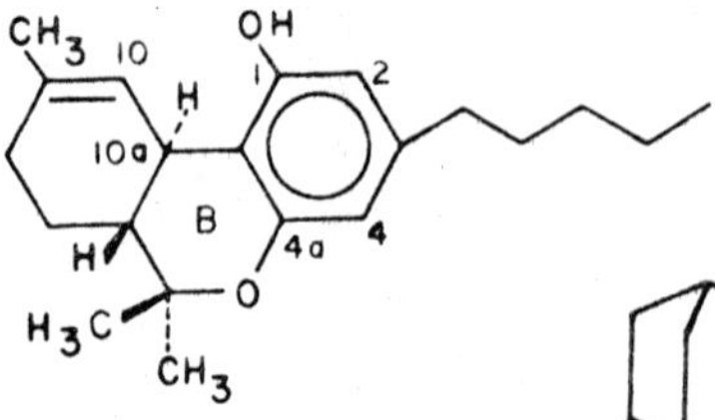

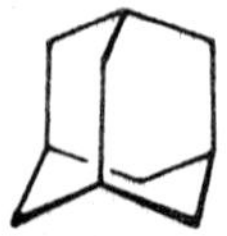

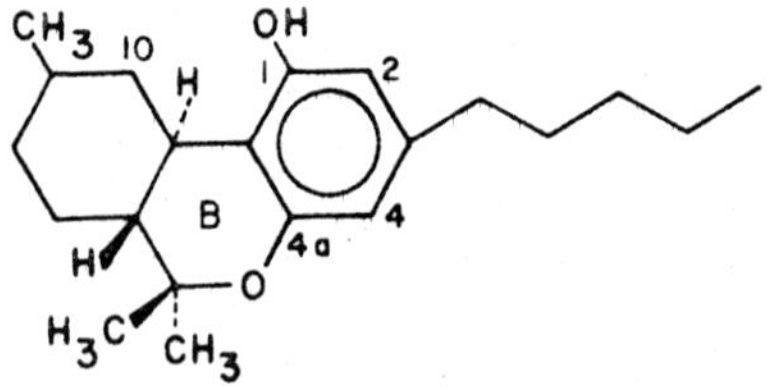

101

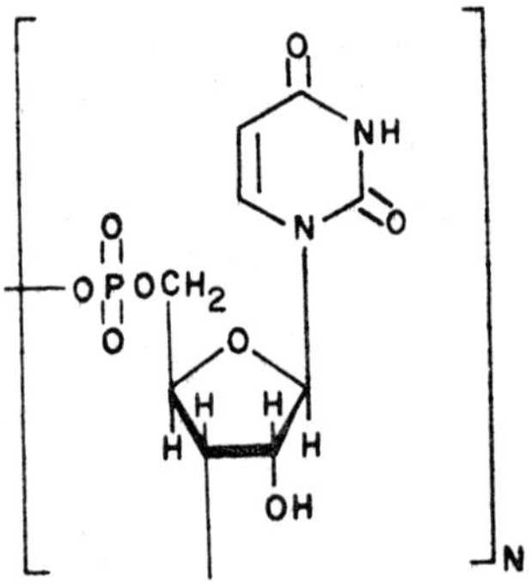

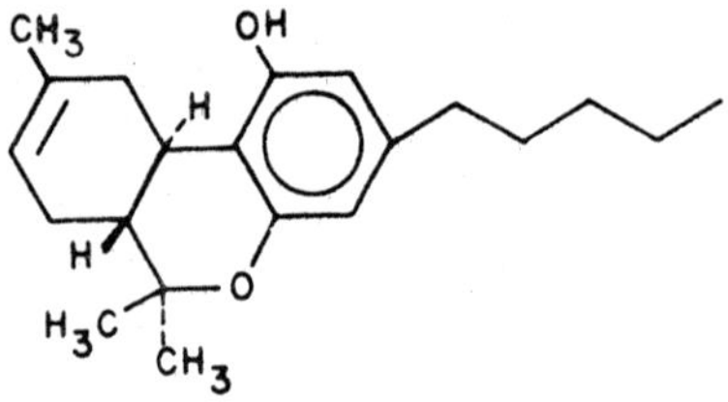

102

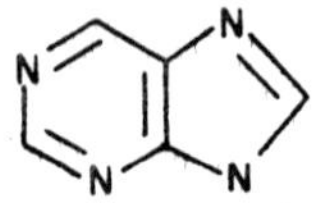

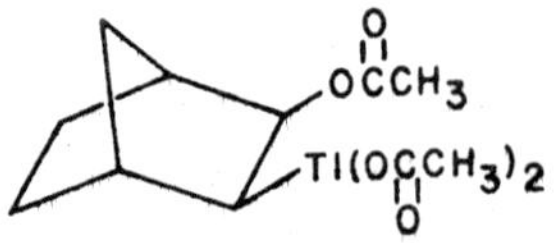

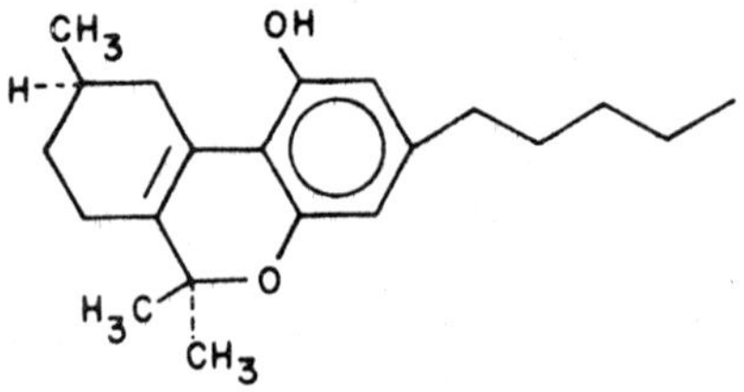

103

104

105

106

107

108

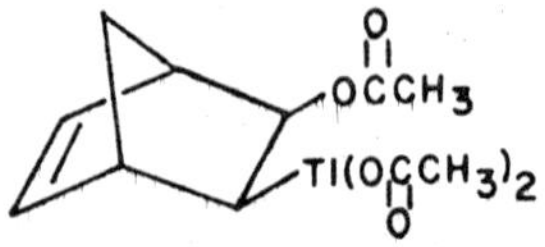

109

REFERENCES

1. F. A. L. Anet and A. J. R. Bourn, *J. Amer. Chem. Soc.* **87**, 5250 (1965).
2. N. S. Bhacca, L. F. Johnson, and J. N. Shoolery, "NMR Spectra Catalog." Varian Assoc., Palo Alto, California, 1961.
3. V. J. Kowalewski and D. G. de Kowalewski, *J. Chem. Phys.* **32**, 1272 (1960).
4. J. V. Hatton and R. E. Richards, *Mol. Phys.* **5**, 139 (1962).
5. L. A. LaPlanche and M. T. Rogers, *J. Amer. Chem. Soc.* **85**, 3728 (1963).
6. F. A. L. Anet, A. J. R. Bourn, P. Carter, and S. Winstein, *J. Amer. Chem. Soc.* **87**, 5249 (1965).
7. M. Ohtsuru, M. Teraoka, K. Tori, and K. Takeda, *J. Chem. Soc. B* p. 1033 (1967).
8. R. H. Martin and J. C. Nouls, *Tetrahedron Lett.* p. 2727 (1968).
9. S. H. Grover and J. B. Stothers, *J. Amer. Chem. Soc.* **91**, 4331 (1969).
10. A. L. Thakkar, L. G. Tensmeyer, R. B. Hermann, and W. L. Wilham, *J. Chem. Soc. D* p. 524 (1970).
11. R. Ottinger, G. Boulvin, J. Riesse, and G. Chiurdoglu, Tetrahedron **21**, 3435 (1965).
12. T. G. Alexander and M. Maienthal, *J. Pharm. Sci.* **53**, 962 (1964).
13. M. W. Hanna and A. Sandoval, *Biochim. Biophys. Acta* **155**, 433 (1968).
14. Y. Takino, A. J. Ferretti, V. Flanagan, M. Gianturco, and M. Vogel, *Tetrahedron Lett.* p. 4019 (1965).
15. A. Holmes, A. G. Brown, C. P. Falshaw, E. Haslam, and W. D. Ollis, *Tetrahedron Lett.* p. 1193 (1966).
16. T. Bryce, P. D. Collier, I. Fowlis, P. E. Thomas, D. Frost, and C. K. Wilkins, *Tetrahedron Lett.* p. 2789 (1970).
17. A. J. Jones, D. M. Grant, and K. F. Kuhlmann, *J. Amer. Chem. Soc.* **91**, 5013 (1969).
18. A. J. Jones, T. D. Alger, D. M. Grant, and W. M. Litchman, *J. Amer. Chem. Soc.* **92**, 2386 (1970).
19. K. Kuhlmann and J. D. Baldeschwieler, *J. Amer. Chem. Soc.* **85**, 1010 (1963).
20. R. Kaiser, *J. Chem. Phys.* **39**, 2435 (1963).
21. J. C. Nouls, G. Van Binst, and R. H. Martin, *Tetrahedron Lett.* 4065 (1967).
22. G. Van Binst, J. C. Nouls, J. Stokoe, C. Danheux, and R. H. Martin, *Bull. Soc. Chim. Belg.* **74**, 506 (1965).
23. J. G. Colson, P. T. Lansbury, and F. D. Saeva, *J. Amer. Chem. Soc.* **89**, 4987 (1967).
24. P. T. Lansbury, A. J. Lacher, and F. D. Saeva, *J. Amer. Chem. Soc.* **89**, 4361 (1967).
25. P. T. Lansbury and F. D. Saeva, *J. Amer. Chem. Soc.* **89**, 1890 (1967).
26. J. C. Nouls, P. Wollast, J. C. Braekman, G. Van Binst, J. Pecher, and R. H. Martin, *Tetrahedron Lett.* p. 2731 (1968).
27. R. W. Franck and K. Yanagi, *J. Org. Chem.* **33**, 811 (1968).
28. N. Abe, R. Onoda, K. Shirahata, T. Kato, M. C. Woods, and Y. Kitihara, *Tetrahedron Lett.* p. 1993 (1968).
29. M. C. Woods, I. Miura, A. Ogiso, M. Kurabayashi, and H. Mishima, *Tetrahedron Lett.* p. 2009 (1968).
30. K. Takeda, I. Horibe, M. Teraoka, and H. Minato, *J. Chem. Soc. D* p. 940 (1968).
31. J. K. Saunders, R. A. Bell, C.-Y. Chen, D. B. MacLean, and R. H. F. Manske, *Can. J. Chem.* **46**, 2876 (1968).
32. D. B. MacLean, R. A. Bell, J. K. Saunders, C.-Y. Chen, and R. H. F. Manske, *Can. J. Chem.* **47**, 3593 (1969).
33. R. A. Bell and J. K. Saunders, *Can. J. Chem.* **46**, 3421 (1968).
34. R. H. F. Manske, R. G. A. Rodrigo, D. B. MacLean, D. E. F. Gracey, and J. K. Saunders, *Can. J. Chem.* **47**, 3589 (1969).

35. R. H. F. Manske, R. Rodrigo, D. B. MacLean, D. E. F. Gracey, and J. K. Saunders, *Can. J. Chem.* **47**, 3585 (1969).

36. K. Isono, K. Asahi, and S. Suzuki, *J. Amer. Chem. Soc.* **91**, 7490 (1969).

37. S. Takeuchi, Y. Ogawa, and H. Yonehara, *Tetrahedron Lett.* p. 2737 (1969).

38. S. Ito, M. Kodama, M. Sunagawa, H. Honma, Y. Hayashi, S. Takahashi, H. Ona, T. Sakan, and T. Takahashi, *Tetrahedron Lett.* p. 2951 (1969).

39. M. Koreeda, N. Harada, and K. Nakanishi, *J. Chem. Soc. D* p. 548 (1969).

40. T. Takemoto, S. Arihara, and H. Hikino, *Tetrahedron Lett.* p. 4199 (1968).

41. J. Lugtenburg and E. Havinga, *Tetrahedron Lett.* p. 1505 (1969).

42. R. A. Bell and M. B. Gravestock, *Can. J. Chem.* **47**, 3661 (1969).

43. R. H. Bible and R. R. Burtner, *J. Org. Chem.* **26**, 1174 (1961).

44. T. Tomimatsu, M. Hashimoto, T. Shingu, and K. Tori, *J. Chem. Soc. D* p. 168 (1969).

45. D. J. Aberhart, K. H. Overton, and S. Hunek, *J. Chem. Soc. D* p. 162 (1969).

46. G. R. Newkome and N. S. Bhacca, *J. Chem. Soc. D* p. 385 (1969).

47. R. E. Schirmer, J. H. Noggle, J. P. Davis, and P. A. Hart, *J. Amer. Chem. Soc.* **92**, 3266 (1970); erratum; **92**, 7239 (1970).

48. M. Plat, M. Hachem-Mehri, M. Koch, U. Scheidegger, and P. Potier. *Tetrahedron Lett.* p. 3395 (1970).

49. M. Koreeda and K. Nakanishi, *J. Chem. Soc. D* p. 351 (1970).

50. Y. Nakatani, T. Yamanishi, N. Esumi, and T. Suzuki, *Agr. Biol. Chem.* **34**, 152 (1970).

51. Y. Nakatani and T. Yamanishi, *Tetrahedron Lett.* p. 1955 (1969).

52. A. Sato and H. Mishimi, *Tetrahedron Lett.* p. 1803 (1969).

53a. R. R. Fraser and F. J. Schuber, *Can. J. Chem.* **48**, 633 (1970).

53b. R. F. Watson and J. F. Eastham, *J. Amer. Chem. Soc.* **87**, 664 (1965).

53c. K. Mislow, M. M. Green, P. Laur, and D. R. Chisholm, *J. Amer. Chem. Soc.* **87**, 665 (1965).

54. I. W. J. Still, C. J. MacDonald, and Y.-N. Oh, *Can. J. Chem.* **48**, 1526 (1970).

55. M. C. Woods, I. Miura, Y. Nakadaira, A. Terahara, M. Maruyama, and K. Nakanishi, *Tetrahedron Lett.* **4**, 321 (1967).

56a. M. Maruyama, A. Terahara, Y. Itagaki, and K. Nakanishi, *Tetrahedron Lett.* pp. 299, 303 (1967).

56b. M. Maruyama, A. Terahara, Y. Nakadaira, M. C. Woods, and K. Nakanishi, *Tetrahedron Lett.* p. 309 (1967).

56c. M. Maruyama, A. Terahara, Y. Nakadaira, M. C. Woods, Y. Takagi, and K. Nakanishi, *Tetrahedron Lett.* p. 315 (1967).

57. K. Tori, M. Ohtsuru, I. Horibe, and K. Takeda, *J. Chem. Soc. D* p. 943 (1968).

58. K. Tori and I. Horibe, *Tetrahedron Lett.* p. 2881 (1970).

59. M. C. Woods, H.-C. Chiang, Y. Nakadaira, and K. Nakanishi, *J. Amer. Chem. Soc.* **90**, 522 (1968).

60. R. A. Bell and E. N. C. Osakwe, *J. Chem. Soc. D* p. 1093 (1968).

61. R. A. Archer and P. V. DeMarco, *J. Amer. Chem. Soc.* **91**, 1530 (1969).

62. R. D. G. Cooper, P. V. DeMarco, J. C. Cheng, and N. D. Jones, *J. Amer. Chem. Soc.* **91**, 1408 (1969).

63. R. D. G. Cooper, P. V. DeMarco, and D. O. Spry, *J. Amer, Chem. Soc.* **91**, 1528 (1969).

64. R. D. G. Cooper, P. V. DeMarco, C. F. Murphy, and L. A. Spangle, *J. Chem. Soc. C* p. 340 (1970).

65. N. S. Bhacca and N. H. Fischer, *J. Chem. Soc. D* p. 68 (1969).

66. Y. Ishizaki, Y. Tanahashi, T. Takahashi, and K. Tori, *J. Chem. Soc. D* p. 551 (1969).

67. J. Lugtenburg and E. Havinga, *Tetrahedron Lett.* p. 2391 (1969).

68. M. C. Woods, S. Ebine, M. Hoshino, K. Takahashi, and I. Miura, *Tetrahedron Lett.* p. 2879 (1969).

69. A. R. Dunn and R. J. Stoodley, *Tetrahedron Lett.* p. 3367 (1969).

70. P. A. Hart and J. P. Davis, *J. Amer. Chem. Soc.* **91**, 512 (1969).

71. P. A. Hart and J. P. Davis, *Biochem. Biophys. Res. Commun.* **34**, 733 (1969).

72. R. A. Archer, D. B. Boyd, P. V. DeMarco, I. J. Tyminski, and N. L. Allinger, *J. Amer. Chem. Soc.* **92**, 5200 (1970).

73a. I. Solomon, *Phys. Rev.* **99**, 559 (1955).

73b. I. Solomon and N. Bloembergen, *J. Chem. Phys.* **25**, 261 (1956).

74. P. S. Hubbard, *J. Chem. Phys.* **42**, 3546 (1965).

75. J. H. Chaffin, III and P. S. Hubbard, *J. Chem. Phys.* **46**, 1511 (1967).

76. E. L. Mackor and C. Maclean, *J. Chem. Phys.* **42**, 4254 (1965).

77. R. J. C. Brown, H. S. Gutowsky, and K. Shimomura, *J. Chem. Phys.* **38**, 76 (1963).

78. K. F. Kuhlmann, D. M. Grant, and R. K. Harris, *J. Chem. Phys.* **52**, 3439 (1970).

79. T. Fukumi, Y. Arata, and S. Fujiwara, *J. Mol. Spectros.* **27**, 443 (1968).

80. R. Kaiser, *J. Chem. Phys.* **42**, 1838 (1965).

81. S. I. Chan and G. P. Kreishman, *J. Amer. Chem. Soc.* **92**, 1102 (1970).

82. E. G. Paul and D. M. Grant, *J. Amer. Chem. Soc.* **86**, 2977 (1964).

83. K. F. Kuhlmann and D. M. Grant, *J. Amer. Chem. Soc.* **90**, 7355 (1968).

84. J. Feeney, D. Shaw, and P. J. S. Pauwels, *J. Chem. Soc. D* p. 554 (1970).

85. A. Allerhand and D. W. Cochran, *J. Amer. Chem. Soc.* **92**, 4482 (1970).

86. F. A. L. Anet, *Tetrahedron Lett.* p. 3399 (1964).

87. J. V. Hatton, *J. Chem. Phys.* **40**, 933 (1964).

88. R. A. Bell and J. K. Saunders, *J. Chem. Soc. D* p. 1078 (1970).

89. D. F. S. Natusch and R. E. Richards, *J. Chem. Soc. D* p. 185 (1966).

90. R. A. Bell and J. K. Saunders, *Can. J. Chem.* **48**, 1114 (1970).

A P P E N D I X

I

TIGHTLY COUPLED SPINS

In the preceding chapters we have explicitly and invariably assumed that all of the nuclear spins involved were loosely coupled, with their scalar coupling constant J being much less than their relative chemical shift δ. We shall now give a brief discussion of tightly coupled spins in order to (a) find how and why tight coupling alters the simple theory previously presented and (b) estimating the practical limit on the J/δ ratio which will permit the simple theory of the previous chapters to be used accurately.

First, how do tightly coupled spins differ from loosely coupled spins. For two loosely coupled spins-$\frac{1}{2}$ the eigenfunctions of the combined spins were simple products of the eigenfunctions of the separated spins: i.e., $\alpha\alpha$, $\alpha\beta$, $\beta\alpha$. and $\beta\beta$. Likewise, the allowed transitions of the system could be identified as the transitions of one or the other of the component spins: e.g., the transition $\alpha\beta \rightarrow \alpha\alpha$ is a spin flip of spin 2 and the transition $\beta\alpha \rightarrow \alpha\alpha$ is a spin flip of spin 1. When $J \sim \delta$, the off-diagonal terms of the $J\mathbf{I}_1 \cdot \mathbf{I}_2$ coupling mix these functions and the eigenfunctions are*:

$$\psi_1 = \alpha\alpha \tag{I.1a}$$

---

* This subject is treated in nearly all books on NMR. Here we follow the conventions and notation of Carrington and McLachlan (*1*).

226

$$\psi_2 = \cos\theta\,\alpha\beta - \sin\theta\,\beta\alpha \qquad (\text{I.1b})$$

$$\psi_3 = \sin\theta\,\alpha\beta + \cos\theta\,\beta\alpha \qquad (\text{I.1c})$$

$$\psi_4 = \beta\beta \qquad (\text{I.1d})$$

The coefficients can be obtained from

$$\sin 2\theta = J/(J^2 + \delta^2)^{1/2} \qquad (\text{I.2a})$$

$$\cos 2\theta = \delta/(J^2 + \delta^2)^{1/2} \qquad (\text{I.2b})$$

$$\sin\theta = \left[\frac{1 - \cos 2\theta}{2}\right]^{1/2} \qquad (\text{I.2c})$$

$$\cos\theta = \left[\frac{1 + \cos 2\theta}{2}\right]^{1/2} \qquad (\text{I.2d})$$

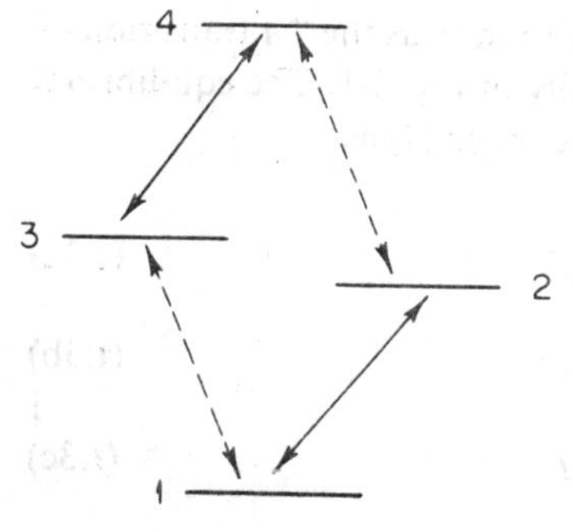

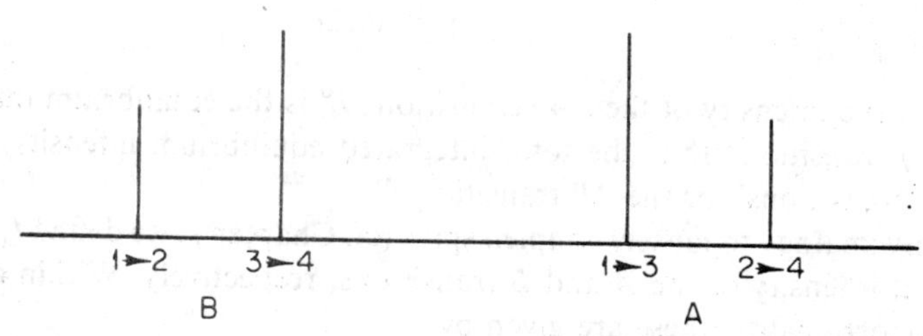

Fig. I-1 The energy level diagram and NMR spectrum of two coupled spins-$\frac{1}{2}$. The **B** transitions are indicated as solid lines and the **A** transitions as dashed lines. Strictly speaking none of the transitions belong exclusively to either spin but for convenience we maintain the conventional labeling used in loosely coupled spin systems.

None of the allowed transitions of this system (Fig. I-1) can be attributed to either of the spins exclusively, e.g., $\psi_1 \rightarrow \psi_1$ is a mixture of $\alpha\beta \rightarrow \alpha\alpha$ and $\beta\alpha \rightarrow \alpha\alpha$, except in the limit $J \ll \delta (\theta \rightarrow 0°)$. Thus, while we can saturate or observe any combination of the four allowed transitions, it is no longer accurate to say that we are "saturating spin 1" and "observing spin 2."

In the extreme limit of tight coupling, $\delta \rightarrow 0$ $(\theta \rightarrow 45°)$ the wave functions ψ_2 and ψ_3 become

$$\psi_2 = (\alpha\beta - \beta\alpha)/2^{1/2}$$

$$\psi_3 = (\alpha\beta + \beta\alpha)/2^{1/2}$$

which are the proper antisymmetric and symmetric functions appropriate for two "identical" spins-$\frac{1}{2}$.

In order to estimate the restriction placed on the J/δ ratio by the term "loose coupling" we shall discuss two spins-$\frac{1}{2}$ in terms appropriate for loosely coupled spins while treating the system mathematically as tightly coupled. Thus, two of the transitions of an AB system, $\psi_2 \rightarrow \psi_1$ and $\psi_4 \rightarrow \psi_3$, are mostly due to spin 2 (or B) if $\cos\theta > \sin\theta$ and will be referred to as "B transitions"; likewise the $\psi_3 \rightarrow \psi_1$ and $\psi_4 \rightarrow \psi_2$ transitions which are primarily (but not entirely) due to spin 1 (or A) will be referred to as the "A transitions."

A typical AB spectrum is shown schematically in Fig. I-1. The equilibrium intensities of the allowed transitions can be calculated (1) as:

B transitions:

$$I^°_{1,2} = \tfrac{1}{2}(1 - S)\, I^° \tag{I.3a}$$

$$I^°_{3,4} = \tfrac{1}{2}(1 + S)\, I^° \tag{I.3b}$$

A transitions:

$$I^°_{1,3} = \tfrac{1}{2}(1 + S)\, I^° \tag{I.3c}$$

$$I^°_{2,4} = \tfrac{1}{2}(1 - S)\, I^° \tag{I.3d}$$

where henceforth

$$S \equiv \sin 2\theta$$

I_{ij} is the intensity of the $i \rightarrow j$ transition; $I^°_{ij}$ is the equilibrium intensity of the $i \rightarrow j$ transition; $I^°$ is the total integrated equilibrium intensity of either the "A transitions" or the "B transitions."

By analogy to loosely coupled spins (cf. Chapter 1) we define I_A and I_B as the total intensity of the A and B transitions, respectively. Within a constant of proportionality, these are given by

$$I_A = I_{1,3} + I_{2,4} = \tfrac{1}{2}(P_1 + P_2 - P_3 - P_4) \tag{I.4a}$$

$$I_B = I_{1,2} + I_{3,4} = \tfrac{1}{2}(P_1 - P_2 + P_3 - P_4) \tag{I.4b}$$

with P_i as usual, the population of the ith energy level. At equilibrium, I° is similarly related to the P_i°.

Using the master equation for populations [Eq. (1.9)], we obtain [compare Eq. (1.13)]:

$$dI_A/dt = -(W_{13} + W_{14})(P_1 - P_1^\circ) - (W_{23} + W_{24})(P_2 - P_2^\circ)$$

$$+ (W_{13} + W_{23})(P_3 - P_3^\circ) + (W_{14} + W_{24})(P_4 - P_4^\circ) \tag{I.5}$$

In Chapter 1 we were able to simplify this equation by assuming that $W_{13} = W_{24} = W_{1A}$, the single quantum transition probability of spin A. For tightly coupled spins, this is no longer possible. In order to keep our treatment as close as possible to that used for tightly coupled spins, we define

$$X \equiv \tfrac{1}{2}(W_{13} + W_{24}) \tag{I.6a}$$

$$Y \equiv \tfrac{1}{2}(W_{13} - W_{24}) \tag{I.6b}$$

Using these definitions, Eq. (I.5) can be rearranged to give

$$dI_A/dt = -X[\Pi_1 + \Pi_2 - \Pi_3 - \Pi_4] - Y[\Pi_1 - \Pi_2 - \Pi_3 + \Pi_4]$$

$$- W_{14}[\Pi_1 - \Pi_4] - W_{23}[\Pi_2 - \Pi_3] \tag{I.7}$$

with $\Pi_i \equiv P_i - P_i^\circ$. We note the following:

$$P_1 - P_4 = I_A + I_B$$

$$P_2 - P_3 = I_A - I_B$$

$$P_1 - P_2 - P_3 + P_4 = 2[I_{1,2} - I_{3,4}]$$

the difference of the intensities of the two B transitions, and

$$P_1^\circ - P_2^\circ - P_3^\circ + P_4^\circ = 2[I_{1,2}^\circ - I_{3,4}^\circ] = -2SI^\circ$$

from Eq. (I.3). Using these identities, Eq. (I.7) becomes

$$dI_A/dt = -(2X + W_{14} + W_{23})(I_A - I^\circ) - (W_{14} - W_{23})(I_B - I^\circ)$$

$$- 2Y(I_{1,2} - I_{3,4}) - 2SYI^\circ \tag{I.8}$$

We now calculate the NOE when the two "B transitions" are saturated and the two "*A* transitions" are observed. Setting

$$I_B = I_{1,2} = I_{3,4} = 0$$

and making a steady-state approximation for I_A, we get

$$f_A(B) = (I_A - I^\circ)/I^\circ = \frac{W_{14} - W_{23} - 2SY}{2X + W_{14} + W_{23}} \tag{I.9}$$

The relaxation of two tightly coupled spins has been considered by a number of authors including, most recently, Freeman *et al.* (2).

For intramolecular dipole–dipole relaxation (dd), the *AB* transition probabilities can be conveniently given in terms of the transition probabilities of loosely coupled spins, W_0, W_{1A}, and W_2 as given by Eqs. (2.9), (2.8), and (2.10), respectively, as (2, 3)

$$W_{13}^{dd} = W_{1A}^{dd}(1 + S) \tag{I.10a}$$

$$W_{24}^{dd} = W_{1A}^{dd}(1 - S) \tag{I.10b}$$

$$W_{23}^{dd} = W_0^{dd}(1 - S^2) \tag{I.10c}$$

$$W_{14}^{dd} = W_2^{dd} \tag{I.10d}$$

Using these we can calculate the NOE for intramolecular dipole–dipole relaxation as (dd only!)

$$f_A(B) = \frac{(W_2^{dd} - W_0^{dd}) - S^2(2W_{1A}^{dd} - W_0^{dd})}{(W_2^{dd} + 2W_{1A}^{dd} + W_0^{dd}) - S^2 W_0^{dd}} \tag{I.11}$$

which reduces to Eq. (1.23) when $S \to 0$. In the extreme narrowing limit we have (Chapter 2)

$$W_0^{dd} : W_{1A}^{dd} : W_2^{dd} = 2 : 3 : 12 \tag{I.12}$$

which gives (dd only!)

$$f_A(B) = \tfrac{1}{2}(1 - 0.4S^2)/(1 - 0.1S^2) \tag{I.13}$$

For loose coupling, of course, the value is 0.5. The correction is likely to be insignificant under coupling conditions which are loose enough for the experiment to be practical. The situation is altered, however, if another mechanism can contribute to the relaxation of the spins.

The only other relaxation mechanisms which have been treated for tightly coupled two spin systems is "relaxation by random external fields." This mechanism can serve as a simplified model for intermolecular relaxation and perhaps for other mechanisms which contribute only to the direct relaxation (i.e., ρ not σ). Noggle (*3*) assumed that the two spins saw random external fields which were, on the average, equal. Freeman *et al.* (*2*) on the basis of more complete data claimed that the presumption of equal external fields was unrealistic and worked out the theory for two spins in the more general case of unequal external fields. The latter case is very complicated so, for the purposes of illustration, we shall use the formulas of Noggle (*3*).

We include this mechanism using a parameter $W_1{}^x$ which, in the loose-coupling limit, would reduce to the single quantum transition probability of the spins due to external relaxation. In this limit it is

$$W_1{}^x = \tfrac{1}{2}\rho^* \qquad (AX \text{ limit}) \tag{I.14}$$

with ρ^* being, perhaps, due to intermolecular dipole relaxation [Eq. (2.29)].

The quantities needed in Eq. (I.9) are

$$W_{13} = W_{1A}^{dd}(1 + S) + W_1{}^x(1 + CS) \tag{I.15a}$$

$$W_{24} = W_{1A}^{dd}(1 - S) + W_1{}^x(1 - CS) \tag{I.15b}$$

$$X = W_{1A}^{dd} + W_1{}^x \tag{I.15c}$$

$$Y = S[W_{1A}^{dd} + CW_1{}^x] \tag{I.15d}$$

$$W_{23} = W_0^{dd}(1 - S^2) + (1 - C)S^2 W_1{}^x \tag{I.15e}$$

$$W_{14} = W_2^{dd} \tag{I.15f}$$

Note that the random external fields mechanism does not contribute to W_{14} and contributes to W_{23} only because of the mixing of the $\alpha\beta$ and $\beta\alpha$ wave functions.

The quantity C in Eqs. (I.15) is the "correlation factor"; it is a measure of the degree to which the two spins *instantaneously* see the same external fields as opposed to seeing the same fields only on the average (*2, 3*). For intermolecular dipole relaxation, it depends on the ratio of the AB internuclear distance to the distance of closest approach of the external dipoles (*3*). For 2,3-dibromothiophene, Noggle (*3*) found $C = 0.75$ and Freeman *et al.* (*2*) found $C = 0.67$. Similar values were found in several other compounds (*2*); smaller values are expected when the AB protons are more distant than in these compounds (*3*).

With relaxation due to random external fields included along with intra-molecular dipole–dipole relaxation the NOE becomes

$$f_A(B) = \frac{\sigma^{dd} + S^2[W_0^{dd} - W_1^{\ x} - 2W_{1A}^{dd} - CW_1^{\ x}]}{\rho^{dd} + \rho^* - S^2[W_0^{dd} - W_1^{\ x} + CW_1^{\ x}]} \tag{I.16}$$

where, by analogy to the loosely coupled case, we have defined

$$\rho^{dd} = W_2^{dd} + 2W_{1A}^{dd} + W_0^{dd} \tag{I.17a}$$

$$\sigma^{dd} = W_2^{dd} - W_0^{dd} \tag{I.17b}$$

$$\rho^* = 2W_1^{\ x} \tag{I.17c}$$

It is difficult to make a general statement about the correction terms in Eq. (I.16) since their effect depends strongly on the ratio ρ^*/ρ^{dd}.

When $C \sim 1$ and $\rho^* \sim 4\rho$, the calculated NOE is, for $J:\delta = 1:6$,

$$f_A(B) \simeq 0.08$$

compared to 0.10 expected in the loose coupling case. With $J:\delta = 1:10$

$$f_A(B) \simeq 0.09$$

the error here is still significant but will be small compared to the usual experimental error. Larger NOE enhancements (ρ^* smaller compared to ρ^{dd}) will have less deviation from the loose coupling formulas.

We conclude that $J:\delta = 1:10$ is probably a safe upper limit to the size of J in a two-spin system. The much more interesting problem, the effect of tight coupling on distance determinations in three spin **ABX** systems remains unsolved. It is probable that the loose coupling formulas of the preceding chapters will be accurate if $(\delta/J) > 10$.

REFERENCES

1. A. Carrington and A. D. McLachlan, "Introduction to Magnetic Resonance." Harper, New York, 1967.
2. R. Freeman, S. Wittekoek, and R. R. Ernst, *J. Chem. Phys.* **52**, 1529 (1970); and references therein.
3. J. Noggle, *J. Chem. Phys.* **43**, 3304 (1965).

APPENDIX

II

MATHEMATICAL METHODS

The problems of curve fitting, numerical integration, and computing Fourier transforms will be discussed in the following sections. The purpose of these sections is to give the reader who is unfamiliar with these methods some understanding of the principles involved and to provide references for those who wish to pursue these topics in more depth. The discussions will be quite general and, of necessity, incomplete.

A. Fitting Exponential Functions to Experimental Data

1. Graphical Estimation by "Curve Stripping"

The constants appearing in the simple exponential function

$$g(t) = ae^{-bt}$$

are easily determined from a plot of $\ln g(t)$ versus t. The slope of such a plot is $-b$, and the line intersects the ordinate at $g(t) = a$. A similar procedure may be used to estimate the constants when $g(t)$ is the sum of two or more exponentials, even though the graph of $\ln g(t)$ versus t will be curved in this case.

233

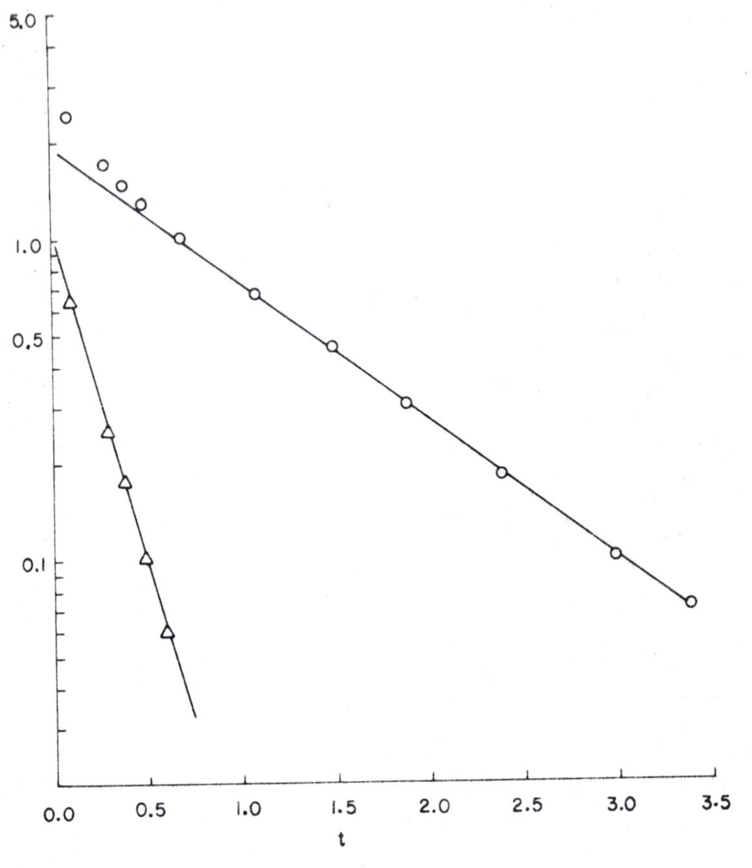

Fig. II-1

The procedure is illustrated in Fig. II-1 for

$$g(t) = a_1 e^{-b_1 t} + a_2 e^{-b_2 t} \qquad\qquad (II.1)$$

Without loss of generality we can assume that b_1 is the smaller of the two exponents. Then a straight line drawn through the reasonably linear portion in the tail of the curve will have a slope of $-b_1$ and will intersect the ordinate at $\ln a_1$. The reason for this is that $b_2 > b_1$ implies that the second term in Eq. (II.1) will decay more rapidly than the first, so that the tail of the curve contains no contribution from the second exponential and thus is given by $\ln g(t) = \ln a_1 - b_1 t$. The remaining parameters a_2 and b_2 are then estimated from the slope and intercept of a plot of $\ln \mathscr{R}$ versus t, where

$$\mathscr{R}(t) = g(t) - a_1 e^{-b_1 t} \equiv a_2 e^{-b_2 t}$$

$$= \text{(experimental } g) - \text{(value of } g \text{ read from the straight line through the tail of the curve)}$$

There are several serious shortcomings of this graphical procedure. The first is that the results can only be considered approximate, and the approximation rapidly becomes worse as the number of exponentials in $g(t)$ increases. A second problem is that the method is restricted to cases where the various exponents are of very different magnitude so that the terms in $g(t)$ decay at very different rates. Finally, taking logarithms in order to linearize exponential data before extracting parameter values from it is not always desirable; the desirability of this procedure depends upon the assumptions one makes about the distribution of errors in the model. Through the use of weighting factors, the numerical least-squares procedures discussed in the following section allow one to incorporate into the calculated parameters any information he has on the relative accuracy of the various experimental values.

2. Least-Squares Estimation

A more precise statement of the problem will be helpful in the following discussion. We have a set of experimental values for the quantity $Y(t)$ as a function of the independent variable t and we wish to describe the data by the model equation $g(\mathbf{a}, t)$. $\mathbf{a}$ is a vector of the parameters entering into $g(\mathbf{a}, t)$, and it is $\mathbf{a}$ that we must determine. In the transient NMR experiments discussed in Chapters 6 and 7, $\mathbf{a}$ would contain all the ρ's, σ's, and I_0's.

The observed values of Y are related to the values of $g(\mathbf{a}, t)$ by

$$Y(t_n) = g(\mathbf{a}, t_n) + e_n \tag{II.2}$$

e_n is an error term which includes differences between Y and g arising from experimental errors, inexactness of the model, or the stochastic nature of Y if Y or t is a random variable. It will be assumed that the number of values of $Y(t_n)$ available exceeds the number of parameters to be determined. The problem then is to find the vector $\mathbf{a}$ which gives the best fit of $g(\mathbf{a}, t)$ to the experimental data.

The next problem is to decide what criterion to use in selecting the "best fit." The most common criterion and the one we shall use here is that of least squares. The vector **a** that gives the "best fit" of $g(\mathbf{a}, t)$ to the $Y(t)$ is taken to be the one which minimizes S, where

$$S = \sum_{n} e_n^2 = \sum [Y(t_n) - g(\mathbf{a}, t_n)]^2 \qquad (\text{II.3})$$

If $g(\mathbf{a}, t)$ is linear in **a**, the least-squares criterion is well understood and has a number of desirable statistical properties. Among these properties is the existence of a single minimum in $S(\mathbf{a})$ as a result of $S(\mathbf{a})$ being quadratic in **a**. When $g(\mathbf{a}, t)$ is not linear in **a** the properties of the least-squares estimates for small sample sizes are not known. Other criteria such as maximum-likelihood or Bayes estimates have occasionally been used and these (or other) measures

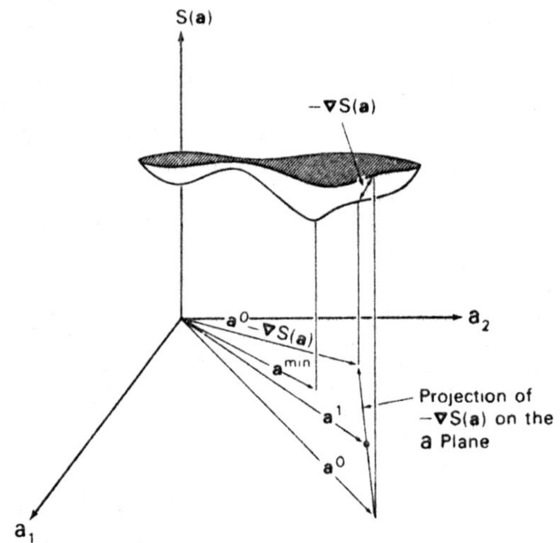

Fig. II-2

of the "goodness of fit" might lead to more useful estimates of the parameters. An additional problem arises in nonlinear cases because there may be more than one minimum in $S(\mathbf{a})$ and the sum of squares surface may be very irregular (see Fig. II-2). Since the nonlinear fitting procedures are all iterative procedures, these irregularities in the surface may result in convergence to the wrong minimum, very slow convergence, or convergence may actually be impossible. A good choice for the initial value of **a** will be very important in obtaining satisfactory convergence. In spite of these problems, the fitting procedures based on the least-squares criterion will remain the most widely used in the absence of progress on other methodology and we will restrict our attention here to the least-squares procedures.

One method which may be used to find the minimum in $S(\mathbf{a})$ is the method of steepest descent. The first step is to select an initial value $\mathbf{a}^0$ for $\mathbf{a}$, perhaps by using the graphical estimation procedure. The negative gradient,

$$-\nabla S(\mathbf{a}^0) = \partial S(\mathbf{a})/\partial \mathbf{a}$$

is then calculated at $\mathbf{a}^0$ using the approximate relation

$$\frac{\partial S(\mathbf{a})}{\partial a_k} \approx \frac{S(\mathbf{a}^0 + \Delta \mathbf{a}_k) - S(\mathbf{a}^0)}{\Delta a_k} \tag{II.4}$$

where $\Delta \mathbf{a}_k$ is a vector with Δa_k as its kth element and zeros for all other elements. $-\nabla S(\mathbf{a}^0)$ is a vector in the direction of most rapid decrease of $S(\mathbf{a})$ with respect to $\mathbf{a}$, so $\mathbf{a}^0$ is moved along $-\nabla S(\mathbf{a}^0)$ to a new value $\mathbf{a}^1$, as is shown in Fig. II-2. This process is repeated until no further decrease in $S(\mathbf{a})$ is obtained. This method often works well on initial iterations, but convergence becomes slow as the minimum is approached. Another disadvantage of this procedure is that convergence is strongly dependent on the scaling of the variables and there is no satisfactory a priori way in which the appropriate scaling can be selected. A complete discussion of these and other aspects of the method of steepest descent can be found in the work of Saaty and Bram (1).

Another common procedure for obtaining least-squares estimates is the Gauss–Newton linearization procedure. The nature of this procedure will be seen most easily if we discuss a linear model first. We begin by writing Eq. (II.2) as the vector equation (II.5)

$$\mathbf{Y}(\mathbf{a}) = \mathbf{g}(\mathbf{a}, t) + \mathbf{e} \tag{II.5}$$

where $\mathbf{Y}$ is a vector of the n observed values, $\mathbf{g}$ is the vector of the n corresponding values calculated from the model, and $\mathbf{e}$ is the vector of errors. If $\mathbf{g}$ were in fact linear in $\mathbf{a}$ we could write

$$\mathbf{g}(\mathbf{a}, t) = \mathbf{G}(t)\mathbf{a} \tag{II.6}$$

where $\mathbf{G}(t)$ is an $n \times N$ matrix, the n rows corresponding to the n observed values and the N columns corresponding to the N parameters in $\mathbf{a}$. Equation (II.3) may be rewritten in our vector notation as

$$S = \mathbf{e}^\mathrm{T} \cdot \mathbf{e} = (\mathbf{Y} - \mathbf{Ga})^\mathrm{T}(\mathbf{Y} - \mathbf{Ga}) \tag{II.7}$$

where the superscript T denotes the transpose matrix. The minimum value of S may be found by setting $\partial S/\partial \mathbf{a}$ equal to zero, which gives

$$0 = -\mathbf{G}^\mathrm{T}(\mathbf{Y} - \mathbf{Ga}) \tag{II.8}$$

We have used the relations

$$\frac{\partial}{\partial x} \mathbf{V}(x)^{\mathrm{T}} \cdot \mathbf{V}(x) = 2 \frac{\partial \mathbf{V}(x)^{\mathrm{T}}}{\partial x} \cdot \mathbf{V}(x) = 2\mathbf{V}(x)^{\mathrm{T}} \cdot \frac{\partial \mathbf{V}(x)}{\partial x}$$

and

$$\partial \mathbf{Y}/\partial \mathbf{a} = 0$$

in obtaining (II.8). Equation (II.8) is easily solved for $\mathbf{a}$ and gives

$$\mathbf{a} = (\mathbf{G}^{\mathrm{T}}\mathbf{G})^{-1}\mathbf{G}^{\mathrm{T}}\mathbf{Y} \qquad\qquad\qquad (\mathrm{II.9})$$

where a superscript -1 denotes the inverse of the matrix. When $\mathbf{g}$ is linear in $\mathbf{a}$, the least-squares value of $\mathbf{a}$ is given by (II.9). When $\mathbf{g}$ is not linear in $\mathbf{a}$, we first perform a Taylor expansion of $\mathbf{g}(\mathbf{a}, t)$ about $\mathbf{a}^0$ and, keeping only the linear terms in the expansion, proceed as we did in deriving (II.9).

The Taylor series for $\mathbf{g}$ about $\mathbf{a}^0$ may be written

$$\mathbf{g}(\mathbf{a}, t) = \mathbf{g}(\mathbf{a}^0, t) + \frac{\partial \mathbf{g}(\mathbf{a}^0, t)}{\partial \mathbf{a}} \cdot (\mathbf{a} - \mathbf{a}^0) + \cdots \qquad\qquad (\mathrm{II.10})$$

Keeping only the terms shown in (II.10) we may write

$$\mathbf{e} = (\mathbf{Y} - \mathbf{g}) = [\mathbf{Y} - \mathbf{g}(\mathbf{a}^0, t) - \mathbf{G}(\mathbf{a}^0, t)(\mathbf{a} - \mathbf{a}^0)]$$

where $\mathbf{G}(\mathbf{a}^0, t) = \partial \mathbf{g}(\mathbf{a}^0, t)/\partial \mathbf{a}$. In practice, these derivatives are calculated using Eq. (II.4). Comparing this with (II.7) and (II.9) we expect the minimum in the sum of squares to be approximated by $\mathbf{a}^1$ where

$$\mathbf{a}^1 - \mathbf{a}^0 = (\mathbf{G}^{\mathrm{T}}\mathbf{G})^{-1}\mathbf{G}^{\mathrm{T}}[\mathbf{Y} - \mathbf{g}(\mathbf{a}^0, t)] \qquad\qquad (\mathrm{II.11})$$

The fact that we have approximated g by the first terms in its expansion about $\mathbf{a}^0$ means that $\mathbf{a}^1$ will only approximate the vector of constants that gives the true minimum in S, but it is, hopefully, a better approximation than $\mathbf{a}^0$ was. This procedure may now be iterated, expanding next about $\mathbf{a}^1$, then about $\mathbf{a}^2$, and so on until the minimum in S is located with the desired accuracy. Additional discussion of this method may be found in the work of Booth *et al.* (2). Let it suffice to say that the procedure often works well when $\mathbf{a}^k$ is near the minimum.

Hartley (3) has described a modified form of the Gauss–Newton method in which convergence is assured, and may be hastened in some cases. Another modification has been suggested by Jennrich and Sampson (4) which is intended to avoid the difficulties arising in the Gauss–Newton method due to the high correlations which often exist among the parameter estimates, causing the matrix $G^T G$ to be nearly singular.

A method due to Marquardt (5) which combines the linearization procedure of the Gauss–Newton method with the method of steepest descent has found wide application in nonlinear problems. In this method the correction vector is calculated using

$$\mathbf{a} - \mathbf{a}^0 = (G^T G + \lambda I)^{-1} G^T [Y - g(\mathbf{a}^0, t)] \qquad \text{(II.12)}$$

in place of Eq. (II.11). λ is a nonnegative number and I is the identity matrix. If γ is the angle between $(\mathbf{a} - \mathbf{a}^0)$ given by equation (II.12) and the value of $(\mathbf{a} - \mathbf{a}^0)$ that one would obtain from the method of steepest descent, then $\gamma \to 0$ monotonically as $\lambda \to \infty$. We also see that (II.12) will approach the Gauss–Newton correction Eq. [(II.11)] as $\lambda \to 0$. A computation using Marquardt's procedure uses a large value of λ on the initial iterations to take advantage of the smooth convergence expected from the method of steepest descent when we are far from the minimum in $S(\mathbf{a})$. As iteration proceeds, λ is systematically decreased so that the procedure approaches the Gauss–Newton procedure as the minimum in the sum of squares surface is approached. Thus Marquardt's algorithm should share with the method of steepest descent the ability to converge rapidly from a region far from the minimum, and yet should converge as rapidly as the Gauss–Newton method once the region of the minimum has been entered. For a more detailed discussion, including the way in which scaling is introduced into Eq. (II.12), the reader is referred to Marquardt's paper (5).

B. Numerical Integration

The numerical integration problem is the problem of estimating I, where

$$I = \int_a^b f(x)\, dx$$

given only the values of $f(x)$ at selected points. The most common procedures are based upon approximating $f(x)$ by an nth order polynomial, $P_n(x)$, selected to pass through the known values of $f(x)$. The approximation need only be good, of course, on the interval $[a, b]$, or subintervals of $[a, b]$. The common integration procedures using the polynomial approximation may be divided into two groups: the Newton–Cotes formulas that use values of $f(x)$ at equally spaced points; and the Gaussian formulas that use values at selected, unequally spaced points.

On the basis of the accuracy required, and the simplicity of the various methods, the Newton–Cotes procedure with $n = 1$ or $n = 2$ is the method of choice for the calculations discussed in Chapter 4. With $n = 1$, the polynomial describes a straight line and the method is better known as the trapezoidal rule. With $n = 2$, the polynomial is quadratic and we have Simpson's rule. We will limit ourselves to a brief discussion of these two methods. The notation to be used is defined in Fig. II-3 where the interval to be integrated, $[a, b]$, has been divided into $N - 1$ segments of equal length h.

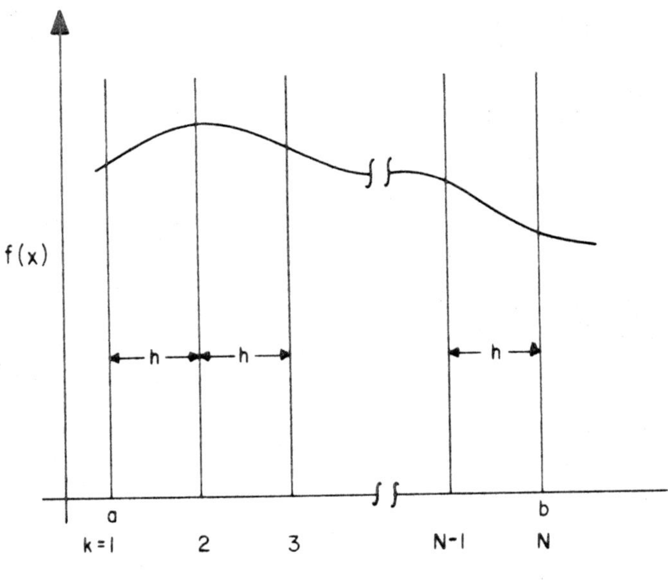

Fig. II-3

For $n = 1$ (trapezoidal rule), the values of $f(x)$ between the known values are estimated by drawing straight line segments between f_1 and f_2, f_2, and f_3, etc. The area under the kth segment is $\frac{1}{2}[f(x_k) + f(x_{k+1})]$, so the integral I is approximated by

$$I \doteq \tfrac{1}{2}[f(x_1) + f(x_N)] + \sum_{i=2}^{N-1} f(x_i)$$

The error in this approximation can be shown to be

$$-(h^3/12)(N - 1)f''(\zeta)$$

where $f''(\zeta)$ is the second derivative of $f(x)$ with respect to x evaluated at some point ζ on $[a, b]$. The magnitude of the error cannot exceed $h^3(N-1)|f''|_{max}$ where $|f''|_{max}$ is evaluated on $[a, b]$. We see that, as expected, the estimation of I becomes exact if $f(x)$ is a straight line so that $f''(x) = 0$.

Simpson's rule uses a quadratic ($n = 2$) interpolation between the known values of $f(x)$ rather than a linear interpolation. It also requires $N-1$ to be even, i.e., that $[a, b]$ be divided into an even number of subintervals. The value of I is then estimated by

$$I = (h/3)\,[f(x_1) + 4f(x_{N-1}) + f(x_N)] + (h/3)\left[\sum_{i=1}^{(N-3)/2}[2f(2i+1) + 4f(2i)]\right]$$

The error term can be written (9) as

$$-(N-1)(h^5/90)\,f^{(4)}(\zeta)$$

where $f^{(4)}(\zeta)$ is the fourth derivative of $f(x)$ with respect to x and ζ is some value of x on the interval $[a, b]$. Again, an upper limit on the error can be specified

$$|\text{error}| \leqslant (N-1)(h^5/90)\,|f^{(4)}|_{max}$$

The dependence of the error on the fourth derivative is unexpected as the error terms typically depend on the $(n+1)$th derivative. It is, in fact, the restriction to an even number of intervals which causes the third-order derivatives to drop out of the error expression for Simpson's rule. That the resulting accuracy should be quite sufficient for the problems discussed in Chapter 4 can be seen by looking at the plots of enhancement versus θ in Fig. 4-4. The changes in slope are gradual and one would anticipate very small values for the error providing $P(\theta)$ is equally well behaved.

C. The Fourier Transform

The complex Fourier transform of a function of time $f(t)$ is defined by

$$F(\omega) = \int_{-\infty}^{+\infty} f(t)\,e^{-i\omega t}\,dt \tag{II.13}$$

The inverse relation is

$$f(t) = (1/2\pi)\int_{-\infty}^{+\infty} F(\omega)\,e^{+i\omega t}\,dt \tag{II.14}$$

Using $e^{-i\omega t} = \cos\omega t + i\sin\omega t$, Eq. (II.13) may be rewritten as

$$F(\omega) = \int_{-\infty}^{+\infty} f(t)\cos\omega t\,dt + i\int_{-\infty}^{+\infty} f(t)\sin\omega t\,dt$$

$$\equiv \mathscr{C}(\omega) + i\mathscr{S}(\omega) \tag{II.15}$$

where $\mathscr{C}(\omega)$ and $\mathscr{S}(\omega)$ are the Fourier cosine and sine transforms of $f(t)$, respectively.

In the application of the Fourier transform to a transient NMR experiment, $f(t)$ would represent the observed decay of the NMR signal following the second rf pulse. $f(t)$ will usually be known at a series of N equally spaced times corresponding to the N channels of the time-averaging computer used to collect the data. $\mathscr{C}(\omega)$ is then calculated for each value of ω by computing $f(t_k)\cos\omega t_k$ for all k and performing the integration numerically, perhaps using one of the procedure discussed in Section B. $\mathscr{S}(\omega)$ is calculated in an analogous fashion.

A substantial reduction in computer time can be realized in practice by using the Cooley–Tukey algorithm (*10*) for fast Fourier transform (FFT) in place of a direct numerical integration. The major factor determining the time required to compute a transform is the time required for the multiplications. By taking advantage of symmetries inherent in the transform process, the FFT algorithm drastically reduces the number of multiplications and the amount of time required for the computation.

A very good review of the FFT has been given by Bergland (*11*).

GENERAL REFERENCES

Box, G. E. P., Fitting empirical data. *Ann. N. Y. Acad. Sci.* **86**, 792 (1960).

Box, G. E. P., and Draper, N. R., The Bayesian estimation of common parameters from several responses. Tech. Rep. 31. Dept. of Statistics, Univ. of Wisconson, Madison, Wisconson, 1964.

Box, G. E. P., and Hunter, W. G., A useful method for model building. *Techometrics* **4**, 301 (1962).

Box, G. E. P., and Hunter, W. G., The experimental study of physical mechanisms. *Technometrics* **7**, 23 (1965).

Bracewell, R., "The Fourier Transform and Its Applications." McGraw-Hill, New York, 1965.

Brigham, E. O., and Morrow, R. E., The fast Fourier transform. *IEEE Spectrum* **4**, 63 (1967).

Draper, N. R., and Smith, H., "Applied Regression Analysis," Chapter 10. Wiley, New York, 1966.

Metzler, C. M., A brief introduction to nonlinear least squares estimation. *Compilation Symp. Papers Nat. Meeting APhA Acad. Pharm. Sci., 5th, Washington, D.C. 1968.* Amer. Pharm. Assoc., Washington D.C. 1970.

Simon, W., A method of exponential separation applicable to small computers. *Phys. Med. Biol.* **15**, 355 (1970).

REFERENCES

1. T. L. Saaty and J. Bram, "Nonlinear Mathematics," pp. 70–88. McGraw-Hill, New York, 1964.
2. G. W. Booth, G. E. P. Box, M. E. Muller, and T. I. Peterson, "Forecasting by Generalized Regression Methods, Nonlinear Estimation." I.B.M. Corp., New York, 1959.
3. H. O. Hartley, *Technometrics* **3**, 269 (1961).
4. R. I. Jennrich and P. F. Sampson, *Technometrics* **10**, 63 (1968).
5. D. W. Marquardt, *J. Soc. Ind. Appl. Math.* **2**, 431 (1963).
6. P. S. Davis and P. Rabinowitz, "Numerical Integration." Ginn (Blaisdell), Boston, Massachusetts, 1967.
7. V. I. Krylov, "Approximate Calculation of Integrals." Macmillan, New York, 1962.
8. B. Carnahan, H. A. Luther, and J. O. Wilkes, "Applied Numerical Methods." Wiley, New York, 1969.
9. J. F. Steffenson, "Interpolation." Williams & Wilkins, Baltimore, Maryland, 1927.
10. J. W. Cooley and J. W. Tukey, *Math. Comput.* **19**, 297 (1965).
11. G. D. Bergland, *IEEE Spectrum* **6**, 41 (1969).

BIBLIOGRAPHY

The papers listed in this section are concerned with the nuclear Overhauser effect or contain an application of it to a chemical problem. These papers, which have not been cited or referenced elsewhere in the text, complete the listing of all papers on the NOE which have come to our attention through June, 1971

T. D. Alger, S. W. Collins, and D. M. Grant, Carbon-13 relaxation time measurements in formic acid, *J. Chem. Phys.* **54**, 2820 (1971).

N. S. Bhacca, L. J. Luskus, and K. N. Houk, Elucidation of the structure of the double [6+4] adduct of tropone and dimethylfulvene by nuclear magnetic resonance and the nuclear Overhauser effect, *J. Chem. Soc. D*, 109 (1971).

I. C. Calder, P. J. Garratt, H. C. Longuet-Higgins, F. Sondheimer, and R. Wolovsky, Electrophilic substitution of conjugated eighteen-membered ring systems. Novel conformational effects in monosubstituted [18] annulenes, *J. Chem. Soc. C*, 1041 (1967).

M. Castillo, J. K. Saunders, D. B. MacLean, N. M. Mollov, and G. I. Yakimov, Structure of fumarophycine, *Can. J. Chem.* **49**, 139 (1971).

S. Combrisson, B. Roques, P. Rigny, and J. J. Basselier, Description de l'effet Overhauser dans les systèmes en échange: application à l'analyse conformationnelle d'aldéhydes hétérocycliques, *Can. J. Chem.* **49**, 904 (1971).

D. A. Couch, R. A. DeMario, and J. M. Shreeve, Totally fluorinated esters, $R_F'CO_2R_F^2$, *J. Chem. Soc. D*, 91 (1971).

J. Decazes, J. L. Luche, and H. B. Kagan, Cycloaddition des cetenes sur le bases de Schiff III (I) Determination par R.M.N. de la configuration de triphenyl-1,3,4-alcoyl-3-azetidinones-2, *Tetrahedron Lett.*, 3661 (1970).

E. L. Eliel and F. Nader, Stereochemistry of the reaction of Grignard reagents with ortho esters. A case of orbital overlap control synthesis of unstable polyalkyl-1,3-dioxanes, *J. Amer. Chem. Soc.* **91**, 536 (1969).

B. M. Fung and P. L. Olympia, Jr., ^{19}F double resonance in the presence of chemical exchange, *Mol. Phys.* **19**, 685 (1970).

M. Gordon, W. C. Howell, C. H. Jackson, and J. B. Stothers, Nuclear magnetic resonance study of several derivatives of 1,3,5,7-tetramethyl-tricyclo [$5.1.0.0^{3,5}$] octanes. Stereochemical assignments by nuclear Overhauser enhancements, *Can. J. Chem.* **49**, 143 (1971).

P. A. Hart and J. P. Davis, Pyrimidine nucleoside conformational analysis. Nuclear Overhauser effect and circular dichroism correlations, *J. Amer. Chem. Soc.* **93**, 753 (1971).

R. A. Hoffman and S. Forsén, Transient and steady-state Overhauser experiments in the investigation of relaxation processes. Analogues between chemical exchange and relaxation, *J. Chem. Phys.* **45**, 2049 (1966).

B. Honig, B. Hudson, B. D. Sykes, and M. Karplus, Ring Orientation in β-Ionone and Retinals, *Proc. Nat. Acad. Sci. (U.S.)* **68**, 1289 (1971).

D. S. Kabakoff and E. Namanworth, Nuclear magnetic double resonance studies of the dimethycyclopropylcarbinyl cation. Measurement of the rotation barrier, *J. Amer. Chem. Soc.* **92**, 3234 (1970).

P. D. Kennewell, Applications of the nuclear Overhauser effect in organic chemistry, *J. Chem. Ed.* **47**, 278 (1970).

R. Kosfeld, G. Haegele, and W. Kuchen, Use of the INDOR technique in solution of structural problems in organophosphorus chemistry, *Angew. Chem., Int. Ed. Eng.* **7**, 814 (1968).

A. Kumar and S. L. Gordon, Overhauser studies in two-spin systems, *J. Chem. Phys.* **54**, 3207 (1971).

G. Moraeu, Effet Overhauser application à la chimie organique, *Bull. Soc. Chim. France*, 1770 (1969).

Y. Nakadaira and H. Sakurai, Photochemistry of organolsilicon compounds—1. Photodimerization of 1,1-dimethyl-2,5-diphenyl-1-silacyclopentadiene, *Tetrahedron Lett.*, 1183 (1971).

D. F. S. Natusch, The carbon-13 intensity problem. Elimination of the Overhauser effect with an added paramagnetic species, *J. Amer. Chem. Soc.* **93**, 2566 (1971).

N. Platzer, P. Demerseman, and J.-J. Basselier, Application de l'effet Overhauser nucleaire dans l'etude par RMN de complexes entre le dimethylsulfoxiyde et des phénols substitués, *C. R. Acad. Sci. Paris* **272(C)**, 683 (1971).

O. Sciacovelli, W. von Philipsborn, C. Amith, and D. Ginsburg, An application of homonuclear INDOR spectroscopy. The structure of a dimer of 11,13-dioxo-12-methyl-12-aza [4.4.3] propellane, *Tetrahedron* **26**, 4589 (1970).

M. Sezaki, S. Kondo, K. Maeda, and H. Umezawa, The structure of aquamycin, *Tetrahedron* **26**, 5171 (1970).

S. Takahashi, H. Naganawa, H. Iinuma, T. Takita, K. Maeda, and H. Umezawa, Revised structure and stereochemistry of coriolins, *Tetrahedron Let.*, 1955 (1971).

K. Takeda, I. Horibe, and H. Minato, Preparation of some *cis*-1, *trans*-5 germacratriene derivatives, *J. Chem. Soc.* **D**, 87 (1971).

K. Tori, I. Horibe, K. Kuriyama, and K. Takeda, Conformational isomers and ring inversion of neolinderolactone, a ten-membered-ring furanosesquiterpene, *J. Chem. Soc.* **D**, 957 (1970).

F. W. van Deursen, The detection of hidden proton magnetic resonance signals by means of INDOR, *Org. Mag. Resonance* **3**, 221 (1971).

AUTHOR INDEX

Numbers in parentheses are reference numbers. Numbers in italics indicate the page on which the complete reference is listed.

SUBJECT INDEX